W9-AQS-823

AVIATION
RECORD BREAKERS

AVIATION
RECORD BREAKERS

INNOVATIONS IN MODERN FLIGHT TECHNOLOGY

CHRISTOPHER CHANT

CHARTWELL BOOKS, INC.

A QUINTET BOOK
Published by Chartwell Books
A Division of Book Sales, Inc.
110 Enterprise Avenue
Secaucus, New Jersey 07094

ISBN 1-55521-295-6

This book was designed and produced by
Quintet Publishing Limited
6 Blundell Street
London N7 9BH

Art Director: Peter Bridgewater
Designer: Ian Hunt
Editor: Shaun Barrington

Typeset in Great Britain by
Central Southern Typesetters, Eastbourne
Manufactured in Hong Kong by
Regent Publishing Services Limited
Printed in Hong Kong by
Leefung-Asco Printers Limited

CONTENTS

1

Basic Design and Structure

The late 1980s are seeing dramatic development right across the board of aircraft design. Many of the premises that have been taken for granted for half a century are now being questioned, and a new generation of aircraft is offering completely unforeseen capabilities. Military aircraft are benefiting not in outright performance, which has reached a plateau of what is technically realistic, but rather from features that enhance their tactical capabilities. At the same time, civil aircraft are becoming safer and more economical both to operate and to maintain. The requirements of military and civil aircraft may appear incompatible, but in fact many of the technologies are shared, though in different ways and to different degrees.

The driving force behind the current spate of developments is the military's desire to introduce new generations of combat aircraft. Until the mid-1960s the chief concern of the world's more advanced air arms had been the development and deployment of combat aircraft characterized by extremely high outright performance figures, particularly for speed, rate of climb, service ceiling and range. The first three depended on a clean aerodynamic

McDONNELL DOUGLAS F-4F PHANTOM II

TYPE: *two-seat air-superiority fighter*
WEIGHTS: *empty 30,328 lb/13,757 kg; maximum take-off 61,795 lb/28,030 kg*
DIMENSIONS: *span 38 ft 7½ in/11.77 m; length 63 ft 0 in/19.20 m; height 16 ft 5½ in/5.02 m; wing area 530 sq ft/49.24 m²*
POWERPLANT: *two 17,900-lb/8,119-kg afterburning thrust General Electric J79-GE-17A turbojets*
PERFORMANCE: *speed 1,430 mph/2,301 km/h; ceiling 58,750 ft/17,905 m; range 1,424 miles/2,290 km*
ARMAMENT: *one 20-mm Vulcan multi-barrel cannon and 16,000 lb/7,257 kg of disposable stores (including six Sparrow medium-range or four Sparrow and four Sidewinder short-range air-to-air missles, Maverick air-to-surface missiles, rocket pods and a wide assortment of free-fall bombs) on four special missiles stations plus one underfuselage and four underwing hardpoints*

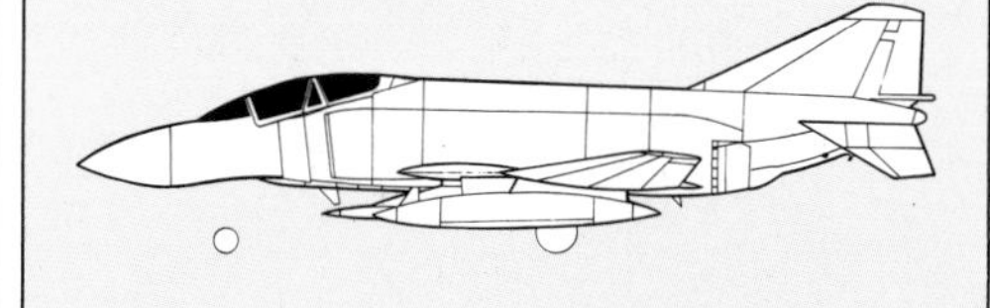

OPPOSITE The greatest Western combat aircraft of the period immediately after World War II, the McDonnell Douglas Phantom II (here the 5,000th aircraft, an F-4E for Turkey) has wholly distinctive aerodynamics and is still a potent warplane some 30 years into its career.

ABOVE Among the earliest Phantom IIs were this pair of F4H-1Fs for the US Navy, who gradually brought them up to definitive F-4B standard.

LEFT The Phantom II also served as an experimental aeroplane in the form of this F-4CCV, fitted with canard foreplanes for enhanced agility, and retaining the experimental fly-by-wire control system pioneered by this particular aeroplane in earlier trials.

LEFT With a few exceptions modern combat aircraft are too expensive for use in a single role, and sights such as this F-4D are common: the aeroplane carries AIM-7 Sparrow air-to-air missiles and laser-guided 'Paveway' guided bombs for a dual, air combat and ground-attack capability.

design combined with a high power-to-weight ratio, and the last on clean aerodynamic design combined with fuel-economical engines and large fuel capacity.

The result was aircraft such as the McDonnell Douglas F-4 Phantom II multi-role fighter with a speed of about Mach 2.25, an initial rate of climb of 30,000 ft/9,145 m per minute, a service ceiling of 60,000+ ft/18,290 m and a typical tactical radius of 700 miles/1,125 km. The subsequent replacement of turbojets by turbofan engines resulted in a useful increase in tactical range on a given fuel load.

The limiting factor on designs of the 1960s and 1970s was the use of aluminium alloys for the primary structure. Such alloys are available in a vast range of tailored varieties, offer considerable strength at modest weight, and do not present great difficulties in airframe manufacture. However, they do lose strength at high temperatures, a factor which becomes critical at airspeeds of about Mach 2.5. Here aerodynamic heating of the airframe begins to present insuperable problems for an aluminium alloy structure. They can be replaced by steel and titanium, but such materials are expensive both to manufacture and to work, and are therefore limited to use in critical high-temperature areas such as wing leading edges and the vicinity of engine nozzles. The only exceptions are specialized aircraft such as the American Lock-

LOCKHEED SR-71 CREW

Appropriately named Blackbird in view of the colour of its airframe's heat-reducing and radar-absorbent coating, the Lockheed SR-71 is an aircraft whose real performance clearly exceeds the figures set by the type in several world record-breaking flights. These include the records for absolute speed in straight-line flight (2,193.17 mph/3,529.56 km/h, established on July 28, 1976), speed over a 1,000-km/621.1-mile closed circuit (2,092.294 mph/3,367.221 km/h on July 27, 1976) and sustained height in horizontal flight (85,059 ft/ 25,929.03 m on July 28, 1976).

The SR-71 operates on the fringes of the atmosphere and the two-man crew – pilot and rear-seat radar systems operator – dress like astronauts: indeed, the S-1010B full-pressure suit designed for the crew of Lockheed U-2 and SR-71 reconnaissance aircraft was used by the astronauts on the first Space Shuttle flights.

While the aircraft is undergoing its time-consuming pre-flight preparation the crew members undergo a full medical examination by the Physiological Support Division attached to each operational unit. It is important that nitrogen be purged from the body to avoid the possibility of the bends, and for about an hour the two men breathe pure oxygen to achieve this end. The men are also given anti-fatigue drugs as a means of reducing muscle stiffness on long flights, and the PSD team check that the crew's special diet has succeeded in reducing bowel activity.

The S-1010B suit is completely sealed, and until the crew members have boarded the aircraft and plugged themselves into the onboard life-support system they have to carry lightweight air-conditioning units to prevent themselves overheating. And at the end of the mission the crew are examined again by the PSD team after being debriefed by the unit intelligence team.

heed SR-71 Blackbird high-altitude strategic reconnaissance aircraft, which has a primary structure of titanium for Mach 3+ performance, and the Soviet Mikoyan MiG-25 Foxbat high-altitude interceptor, which employs a steel primary structure for similar speeds.

Consequently, pending the development of a new structural medium that was both comparatively inexpensive and straightforward to use, mass-production combat aircraft were generally limited to a top speed of Mach 2.5. It was the Americans who first pushed to such a limit in their constant search for tactical aircraft that would have a performance edge over their Soviet counterparts, allied to a technological edge in factors such as radar, other electronic systems and guided missiles.

LOCKHEED SR-71A BLACKBIRD

TYPE: *two-seat strategic reconnaissance aircraft*
WEIGHTS: *empty 60,000 lb/27,216 kg; maximum take-off 170,000 lb/77,111 kg*
DIMENSIONS: *span 55 ft 11 in/16.94 m; length 107 ft 5 in/32.74 m; height 18 ft 6 in/5.64 m; wing area about 1,000 sq ft/92.9 m^2*
POWERPLANT: *two 32,500-lb/14,742-kg afterburning thrust Pratt & Whitney JT11D bleed turbojets*
PERFORMANCE: *speed 2,250 mph/3,620 km/h; ceiling 100,000 ft/30,480 m; range 2,980 miles/4,800 km*
ARMAMENT: *none*

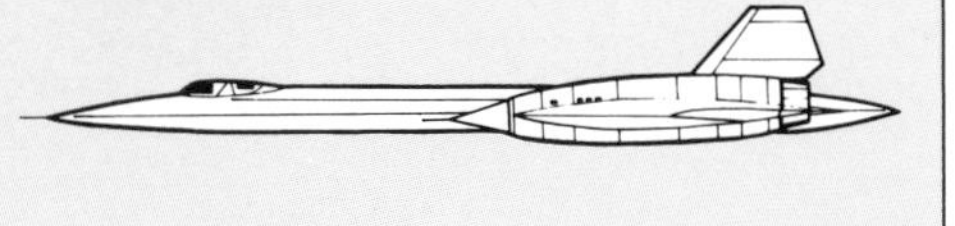

ABOVE Currently the world's ultimate aeroplane, the Lockheed SR-71A strategic reconnaissance platform is highly classified, but holds world records for speed and altitude. The special paint helps to radiate heat and is also radar-absorbent.

MIKOYAN MiG-25 FOXBAT-A

TYPE: *single-seat interceptor*
WEIGHTS: *empty 44,092 lb/20,000 kg; maximum take-off 82,507 lb/37,425 kg*
DIMENSIONS: *span 45 ft 9 in/13.95 m; length 78 ft 1¾ in/23.82 m; height 20 ft 0¼ in/6.10 m; wing area 611.7 sq ft/56.83 m²*
POWERPLANT: *two 27,116-lb/12,300-kg afterburning thrust Tumanskii R-31 turbojets*
PERFORMANCE: *speed 1,849 mph/2,975 km/h; ceiling 80,050 ft/24,400 m; range 1,404 miles/2,260 km*
ARMAMENT: *four AA-6 Acrid long-range air-to-air missiles on underwing hardpoints*

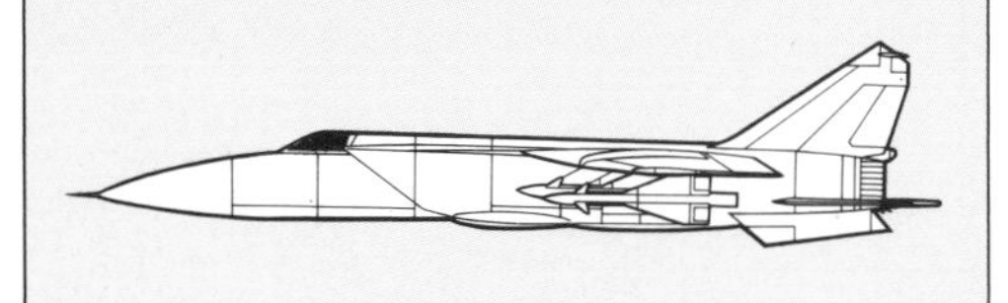

THE VIETNAM EXPERIENCE

The fallacy behind US combat aircraft design was revealed during the Vietnam War, when the US air arms were faced only by the tiny North Vietnamese air force with its small number of obsolescent fighters and large numbers of indifferent surface-to-air missiles, both of Soviet origin. In retrospect it is clear that the surface-to-air missiles posed only a modest threat, though their availability in ever-increasing numbers dictated the development of electronic countermeasures and new tactical gambits. The North Vietnamese fighters also posed only a small threat because of their tiny numbers.

Far more worrying were the fundamental tactical deficiencies revealed when the Americans' latest fighters tackled Soviet-supplied

TOP Successor to the classic MiG-21, the Mikoyan-Gurevich MiG-23 is a capable air-defence and multi-role fighter with modern radar and variable-geometry wings.

ABOVE MIDDLE The MiG-25 has phenomenal performance in a straight line at high altitude, but is less capable against current targets at lower speeds and lower levels. This deficiency is partially remedied in this, the 'Foxbat-E' version.

ABOVE The massive aft-fuselage powerplant installation of the 'Foxbat-E'.

MiG-19 Farmer fighters. The American fighters were designed to use sheer performance to achieve a position where their radars could detect opposing fighters at medium ranges, then engage them with missiles such as the medium-range radar-guided AIM-7 Sparrow or short-range heat-seeking AIM-9 Sidewinder. But the radars of the American fighters proved unreliable and the US rules of engagement dictated that the target must be identified visually before missile launch, ruling out medium-range Sparrow attacks. And at visual ranges the superior tactical capabilities of the Soviet fighters proved a decided embarrassment to the Americans. The MiG-19 is capable of only Mach 1.35, but as the Americans rapidly discovered, any short-range turning engagement almost immediately causes the combatants to lose energy. The resulting drop in speeds toward Mach 0.8 or so meant the better turning rate of the lighter fighter became decisive.

At relatively low speeds and altitudes the Soviet fighters had a decided edge in the real tactical requirements of acceleration, rate of climb and rate of turn. Consequently, in visual combat they were able to outmanoeuvre the larger and heavier American fighters and bring their devastating cannon armament to bear. To make matters worse, the Americans had decided that guns were obsolete, and fighters such as the early Phantoms had been designed without them.

As a short-term expedient podded 20-mm cannon could be carried under the wings or fuselage, but these lacked the installational rigidity of built-in guns and accordingly were less accurate than the Mig-19's three 30-mm weapons.

BELOW Long in the tooth but still a useful weapon and still under further development, the AIM-7 Sparrow is a medium-range missile which homes onto radar energy 'bounced' off the target by the launch fighter's radar.

NAVAL WEAPONS CENTER AIM-9 SIDEWINDER

Designed originally by the Naval Ordnance Test Station (now the Naval Weapons Center) at China Lake, California, and subsequently produced in improved versions by Ford Aerospace and Raytheon, the AIM-9 Sidewinder is the West's most important short-range air-to-air missile. Its development, which encapsulates the whole history of infra-red AAM guidance, began in 1949 using a body of only 5-in/127-mm diameter, and full-scale production was entrusted to Philco Ford (now Ford Aerospace) in 1951.

The first XAAN-N-7 guided prototype was fired in September 1953, and the SW-1 initial production variant reached operational capability in 1956. This model became the AIM-9B in the US forces' tri-service rationalization of designations during 1962. Production reached 80,900 units, and the 155-lb/70.4-kg missile could reach a range of 2 miles/3.2 km with its 20-second mission endurance. The lead sulphide seeker was uncooled, and the claimed 70% single-shot kill probability was attainable only under perfect conditions in a rear-hemisphere attack giving the seeker an unobstructed view of the target's jetpipe.

Subsequent development has resulted in many versions, including the limited-production AIM-9C with semi-active radar homing. The most important developments have concerned: seeker sensitivity for all-aspect engagement capability; greater agility to give dogfighting performance against a manoeuvring target; a more destructive warhead; greater speed; longer range; and a smokeless motor to reduce the give-away plume that characterized all models up to the AIM-9L.

Notable versions were:

- AIM-9D with a nitrogen-cooled seeker, greater speed and range, and a new 22.4-lb/10.2-kg continuous-rod annular blast/fragmentation warhead in place of the original 10-lb/4.54-kg blast type;
- AIM-9E, with a wide-angle seeker;
- AIM-9G, with off-boresight lock-on capability;
- AIM-9H with solid-state electronics;
- AIM-9L with all-aspect seeker capability and greater agility;
- AIM-9M, with greater resistance to countermeasures;
- and AIM-9N with considerably enhanced agility.

The AIM-9R is modelled on the AIM-9N but is more reliable; it weighs 172 lb/78 kg and can reach a range of 11 miles/17.7 km during its 60-second flight-time. The AIM-9 is due to be replaced in the 1990s by the European AIM-132 ASRAAM, but development is continuing.

The Americans moved rapidly to mitigate these shortcomings by enhancing existing aircraft. The F-4E version of the Phantom was produced with a revised nose housing a multi-barrel 20-mm cannon. Subsequently, to improve dogfight agility the outer wing panels were fitted with leading-edge slats. The latter replaced the blown leading-edge flaps incorporated into the original design to reduce landing speeds on carrier decks.

A NEW GENERATION

In the longer term, of course, the Americans saw that they needed a new generation of combat aircraft. The myth of outright superiority in performance had been exploded over Vietnam, and the US services now appreciated what the Soviets had known all along: performance at high altitude was all very well, but surface-to-air missiles had forced operating altitudes down to levels where Mach 2 speeds were not readily attainable and where acceleration, both straight-line and turning, was all-important. Bombers had already been re-cast in the low-level role, in the Boeing B-52 Stratofortress high-altitude strategic bomber being strengthened for its new operating regime and the extraordinary North American B-70 Valkyrie high-altitude Mach-3 bomber – against whose threat the Soviets produced the MiG-25 Foxbat – being cancelled. Now it was the turn of the fighters.

In common with all other aircraft types, the fighter is designed to embody a blend of characteristics optimized for its specific role, whether it be interception, air superiority or air combat. Followed by the Europeans, the Americans now decided that the blend should be revised for greater agility at lower altitudes. Good performance was still needed, but it was clear that the maximization of performance to the detriment of other elements had unbalanced the whole blend.

The new generation of fighters is epitomized by the General Dynamics F-16 Fighting Falcon, winner of a US Air Force competition for a lightweight fighter. The competition was initially seen as a technology demonstration that would evaluate aerodynamic concepts. But by 1975 it was clear that it had produced two prototypes with exceptional combat potential (the General Dynamics YF-16 and Northrop YF-17) and the competition was revised to

LEFT The modern combat aeroplane is increasingly a flying sensor and computing platform to which can be attached an increasing variety of weapons. This is an F-4E armed with an AGM-65 Maverick, a small but highly impressive air-to-surface missile available in variants with different guidance and warhead packages.

RIGHT This F-4E is seen en route to a target in Vietnam during the South-East Asia War, and carries two triplets of 'iron' bombs with nose probes to ensure detonation above the ground.

LEFT This Phantom II's impressive external load includes one drop tank, one electronic countermeasures pod, three Sparrow air-to-air missiles, and two AGM-45 Shrike anti-radar missiles.

ABOVE The North American B-70 Valkyrie supersonic strategic bomber was overtaken by the pace of tactical developments and never entered service, but was in every way a remarkable aircraft of superb performance, and was so highly regarded by the Soviets that they developed the MiG-25 expressly to counter the threat it posed.

ABOVE RIGHT After its termination as a bomber prototype, the B-70 was operated by the US Air Force and NASA as an experimental aeroplane with a number of aerodynamic and structural spin-offs for emerging programmes.

produce a multi-role combat aircraft for the USAF's Tactical Air Command. The rationale behind the F-16 was the desire for a lighter, and therefore cheaper, partner for the massive McDonnell Douglas F-15 Eagle air-superiority fighter. The F-16 entered US service in 1979, and has been adopted subsequently by many of the USA's allies. Meanwhile, the loser in the competition, the YF-17, became the basis for the McDonnell Douglas F/A-18 Hornet dual-role fighter and attack aircraft. The Hornet bears some aerodynamic similarities to the F-16, though it has larger wing/fuselage strakes, twin vertical tail surfaces and twin engines.

CONTROL-CONFIGURED VEHICLES

The YF-16 had been designed from the start as a control-configured vehicle (CCV) of modest performance. A conventional aeroplane is naturally stable, with its centre of gravity located forward of its centre of lift, and requires balance by constant down elevator, resulting in down-load at the tail which is increased considerably in supersonic flight. A CCV, on the other hand, is naturally unstable, with its centre of gravity well aft of its centre of lift: control is effected by powerful aerodynamic surfaces working in association with a computer-directed automatic flight-control system, resulting in up-load at the tail in subsonic flight and only modest down-load in supersonic flight.

The key to CCV design is relaxed static stability, which means that left to itself the airframe would turn end-over-end immediately after takeoff. However, the automatic flight-control system uses its air data system, accelerometers and rate gyros to sense any such tendency immediately, and can then apply corrective control-surface deflections at a rate of more than 100 per second.

Given an automatic flight-control system the weight distribution of the CCV aeroplane can be optimized for agility rather than stability, the computers ensuring that safe flight is maintained and that the pilot's control inputs – on the F-16 these are effected through a small side-stick rather than a centrally-mounted control column – are immediately translated into the optimum control movements. Such a system offers unprecedented levels of agility within the structural limits of the airframe, as well as reducing gust response and structural stress. The last two factors allow a slightly lighter air-frame to be used and extend airframe life.

McDONNELL DOUGLAS F-15A EAGLE

TYPE: *single-seat air-superiority and attack fighter*
WEIGHTS: *empty 27,000 lb/12,247 kg; maximum take-off 68,000 lb/30,845 kg*
DIMENSIONS: *span 42 ft 9¾ in/13.05 m; length 63 ft 9 in/19.43 m; height 18 ft 5½ in/5.63 m; wing area 608 sq ft/56.5 m²*
POWERPLANT: *two 23,950-lb/10,864-kg afterburning thrust Pratt & Whitney F10-PW-100 turbofans*
PERFORMANCE: *speed 1,650+ mph/ 2,655+ km/h; ceiling 60,000 ft/18,290 m; range 2,878+ miles/4,631 km*
ARMAMENT: *one 20-mm Vulcan multi-barrel cannon and up to 16,000 lb/7,258 kg of disposable stores (including four Sparrow medium-range and four Sidewinder short-range air-to-air missiles, guided glide bombs, rocket pods and an exceptionally wide assortment of free-fall conventional or cluster bombs, as well as electronic pods of varying types) on four tangential, three underfuselage and two underwing hardpoints*

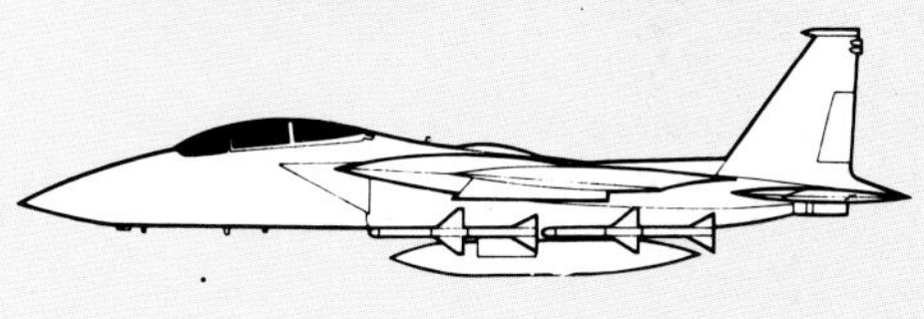

The F-16 is of orthodox configuration, with a single vertical tail surface, mid-set all-moving tailplane and mid-set wings of 40° leading-edge sweep. A closer examination reveals a number of aberrations from the norm, however:

- the tailplane is of the taileron type, the two halves operating in unison for pitch control and differentially for roll control;
- the full-span ailerons are of the flaperon type, again operating differentially for roll control and in unison for increased lift;
- the leading edges are hinged to allow the automatic flight-control system to schedule them as automatic leading-edge flaps and, in concert with the 'flaperons', provide a measure of variable camber to the wing;

McDONNELL DOUGLAS F-15E STRIKE EAGLE

The McDonnell Douglas F-15 Eagle, planned as an air-superiority fighter to succeed the McDonnell Douglas F-4 Phantom II and the older Convair F-106 Delta Dart, began to enter service in November 1974. The type was an immediate success, with a thrust/ weight ratio exceeding unity and consequent excellent climb and high-altitude performance, and proved an admirable attack fighter into the bargain.

The original single-seat F-15A and two-seat F-15B variants – the latter being a combat-capable proficiency trainer – have matured as the F-15C and F-15D. The later versions have upgraded electronics and the ability to carry FAST (Fuel And Sensor Tactical) packs, which attach conformally along the outside of the engine trunks under the wings to increase fuel capacity and electronic capability at minimal cost in weight and drag. The FAST packs are also stressed for the tangential carriage of additional weapons.

The F-15D formed the basis for the F-15E Strike Eagle, an attack-optimized version under development for the US Air force. The first example flew in 1987, and the type offers superb capabilities well suited to the operational requirements of the European theatre. The Strike Eagle was first suggested as the private-venture F-15 Enhanced Eagle, whose two-man crew concept has been used in the Strike Eagle.

The F-15E's main radar, the highly capable Hughes APG-70, is supported by an extensive suite of defensive electronics. The pilot has a wide-angle head-up display, head-down displays and a moving map display, while the rear-seater has no fewer than four head-down displays for weapon management and threat monitoring. The F-15E can also lift a maximum weapon load of 24,250 lb/11,000 kg, compared with the F-15C's 16,000 lb/7,258 kg, can carry all the latest weapons, and is designed to use the podded LANTIRN (Low-Altitude Navigation and Targeting Infra-Red for Night) system for all-weather missions at very low altitudes.

The engine bay is of modular design able to accept either the Pratt & Whitney F100 or the General Electric F110, afterburning turbofans which will eventually deliver a thrust of more than 30,000 lb/13,608 kg. This level of power will allow the F-15E to take off at a weight of 81,000 lb/ 36,741 kg compared with the F-15C's 68,000 lb/30,845 kg.

GENERAL DYNAMICS F-16A FIGHTING FALCON

TYPE: *single-seat air combat and close support fighter*

WEIGHTS: *empty 17,780 lb/8,065 kg; maximum take-off 35,400 lb/16,057 kg*

DIMENSIONS: *span 31 ft 0 in/9.45 m; length 49 ft 4⅞ in/15.09 m; height 16 ft 8½ in/5.09 m; wing area 300 sq ft/27.87 m^2*

POWERPLANT: *one 25,000-lb/11,340-kg afterburning thrust Pratt & Whitney F100-PW-200 turbofan*

PERFORMANCE: *speed 1,320+ mph/2,124+ km/h; ceiling 50,000+ ft/15,240+ m; range 1,150+ miles/1,850+ km*

ARMAMENT: *one 20-mm Vulcan multi-barrel cannon and 20,450 lb/9,276 kg of disposable stores (including six Sidewinder short-range air-to-air missiles, six Maverick air-to-surface missiles or two Harm anti-radar missiles plus a wide assortment of offensive and defensive electronic pods, rocket pods and both guided and unguided free-fall ordnance) on two wingtip missile rails, one underfuselage hardpoint and six underwing hardpoints*

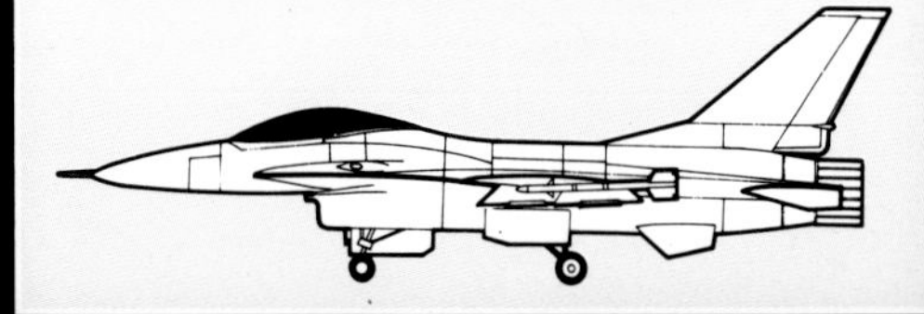

LEFT The second YF-16 prototype on a test flight from the General Dynamics airfield at Fort Worth, Texas, in 1974.

BELOW LEFT The elegant and innovative lines of the first YF16 prototype; the most immediately evident features are the virtually variable-camber wings, the excellent field of vision for the pilot, and the large inlet for engine air under the fuselage.

RIGHT Heavyweight partner to the F-16, the McDonnell Douglas F-15 is designed for air superiority rather than air combat, and possesses quite exceptional performance.

BELOW The F-15 can carry an enormous payload in its secondary attack role, but is best used to secure air mastery with its 20-mm cannon and missile armament of AIM-7 Sparrow medium-range missiles (to be replaced by AIM-120 AMRAAMs) and AIM-9 Sidewinder short-range missiles.

GENERAL DYNAMICS F-16 FIGHTING FALCON IN FOREIGN SERVICE

The General Dynamics F-16 Fighting Falcon was schemed originally as a technology demonstrator for the US Air Force's lightweight fighter competition of 1974. However, its potential as a world-beating combat aircraft was such that it was ordered into production, making its service debut in 1978. Although designed as an air-combat fighter of exceptional agility, it has matured as a superlative all-round combat aircraft equally capable in the attack and close-support and the air-combat roles. Several improved models, both experimental and operational, have been produced and there seems every possibility that the Fighting Falcon will continue to appear in new guises for many years to come.

Given its capabilities, the F-16 has inevitably attracted considerable export interest from US allies and adherents, though the model specifically developed for this role – the F-16/79, with a General Electric J79 turbojet in place of the standard Pratt & Whitney F1100 turbofan – failed to attract significant interest. Even so the basic model has been exported to Belgium, Denmark, Egypt, Greece, Israel, the Netherlands, Norway, Pakistan, Singapore, South Korea, Thailand, Turkey and Venezuela.

Most exports involve standard aircraft, but Israel and Japan are developing upgraded models. Already an F-16 operator when its own IAI Lavi multi-role fighter was cancelled in 1987, Israel has decided to procure additional F-16s incorporating some of the Lavi's systems. Japan has opted for an improved variant with a bigger wing, fully modernized electronics and an element of composite construction. Perhaps most significantly of all, direct-force manoeuvring foreplanes under the inlet trunk, like those pioneered by the F-16/AFTI, will enhance the Japanese Agile Falcon's air-combat and attack manoeuvrability. The bigger wing of the Agile Falcon was originally proposed by General Dynamics to restore the agility lost by the heavier F-16C.

● and the wing is faired into the forward fuselage by highly-swept forebody strakes – often called lerxes, an abbreviation of leading-edge root extensions – to improve the formation of upper-surface vortices and improve handling at high angles of attack.

For practical purposes, therefore, the F-16's wing is of the variable-camber type, which means that it can be shaped in section to suit the five main flight regimes:

● takeoff and landing, with the leading-edge flaps angled up at 2° and the flaperons angled down at 20°;

● initial climb, with the leading-edge flaps angled down at 15° and the flaperons angled down at 20°;

● high-speed flight, with the leading-edge flaps angled up at 2° and the flaperons angled up at 2°;

● manoeuvring, with the leading-edge flaps angled down at 25° and the flaperons level;

● and approach, with the leading-edge flaps down at 15°, flaperons down at 20°.

LEFT The F-16 was designed after the lessons of the South-East Asia War, and was thus not optimized for maximum performance but rather for maximum 'flyability' in the air-combat arena – great agility, high rates of acceleration in all dimensions, and a cockpit offering the pilot unrivalled fields of vision.

The F-16's importance is attested by the number of production lines devoted to its manufacture, firstly in the USA (here) and then in a number of other countries.

The net effect of a fixed central section plus automatically scheduled movable leading- and trailing-edge sections is a wing that can be tailored to the particular conditions at only a modest cost in complexity and weight.

FLY-BY-WIRE CONTROL

The whole system of fixed and moving controls is operated by the F-16's digital flight-control system, often called fly-by-wire. This system interprets inputs from the pilot's force-sensing sidestick controller and rudder pedals to determine his intention. It then produces the optimum combination of control-surface deflections to implement the desired manoeuvre, acting in concert with an automatic aileron/rudder interconnect and yaw-rate limiter. The pilot has a seat inclined at 30° and a single-piece canopy affording excellent fields of view, his semi-reclining position allowing him to withstand higher g forces than is possible in the more conventional upright seat. The HOTAS (hands on throttle and stick) arrangement of

FAR LEFT Propellant gases stream from the muzzle of an F-16's M61A1 Vulcan cannon, buried on the port wing root but still an essential feature for dogfighting air combat at very short range.

LEFT The F-16's warload can include free-fall nuclear weapons, guided glide bombs and air-to-air missiles.

DASSAULT-BREGUET MIRAGE 2000 AND ARMAMENT

A keynote of modern combat aircraft design is flexibility of armament. A typical product of modern practice is the Dassault-Breguet Mirage 2000, which can lift up to 14,209 lb/6,445 kg of disposable stores on its external hardpoints – one centreline unit under the fuselage, two tandem units under each wing root and two units under each wing providing a total of nine weapon-carriage stations.

For air-to-air missions the Mirage 2000 can carry three medium-range Matra Super 530 interception and two short-range Matra R550 Magic dogfighting missiles, plus two internal 30-mm DEFA 554 cannon each supplied with 125 rounds of ammunition. In the air-to-surface role the type is hampered aerodynamically by its large, gust-prone wing, but can carry a wide assortment of useful weapons. Typical ground-attack weapons are a pair of AS.30L laser-homing air-to-surface missiles complemented by a pair of CC421 pods, each carrying a single 30-mm DEFA 554 cannon plus ammunition. Point targets can also be engaged with the Matra laser-guided bomb, an 882-lb/400-kg weapon of which seven can be carried.

Other weapons, less accurate but offering good area-attack potential are:

- the Beluga or Brandt 882-lb/400-kg modular cluster bomb;
- the BAP 100 runway-cratering bomb;
- the BAT 120 anti-vehicle bomb;
- free-fall bombs in various sizes;
- and pods for 2.68-in/68-mm or 3.94-in/100-mm unguided rockets.

The hardpoints can also carry drop tanks or napalm, and at least one is generally used for a podded electronic warfare system.

the cockpit, in which the pilot controls the throttle with his left hand and the sidestick controller with his right hand, enables both arms to be supported, another factor that allows the fly-by-wire system to be fixed at limits of 9 g and 26° angle of attack for conventional operations: it is the pilot's ability to tolerate g forces rather than the structure and/or aerodynamics of the airframe that has now become the limiting factor in air combat.

The European combat aircraft that comes closest to the F-16's CCV concept is the Dassault-Breguet Mirage 2000, which also uses a fly-by-wire control system and wings of simple variable-camber type. In configuration the Mirage 2000 is entirely unlike the F-16, being a tailless delta. Nevertheless, this important French fighter has an altogether higher level of flight agility than its similarly configured predecessor, the Mirage III, since CCV technology allows it to avoid the massive trim drag problems that so eroded the Mirage III's performance in manoeuvring and low-level operations.

MISSION-ADAPTIVE WINGS

Of course, it is better to have a wing that can be tailored over its whole chord to the best shape for any particular set of flight conditions, and such technology is now being pioneered by the mission-adaptive wing (MAW). Designed to be flexed in profile like that of a bird, it has been under development by Boeing since 1979 on the basis of the General Dynamics F-111 strike aircraft, which has the additional bonus of a variable-geometry wing planform.

The trouble with existing variable-camber wings is that the combination of one fixed and two or more movable portions results in dis-

ABOVE The first Dassault-Breguet Mirage 2000B shows off its in-flight refuelling capability.

LEFT One of the keys to the Mirage 2000's capabilities is the combination of a fly-by-wire control system with an advanced wing designed with moveable surfaces on both the leading and trailing edges.

BELOW LEFT A Mirage 2000 shows off part of its external load, in the form of two 374-Imp gal/1700-litre drop tanks, two Magic air-to-air missiles and eight 551-lb/250-kg free-fall bombs.

GENERAL DYNAMICS FB-111A

TYPE: two-seat medium-range operational/strategic bomber with variable-geometry wings
WEIGHTS: maximum take-off 114,300 lb/51,846 kg
DIMENSIONS: span 70 ft 0 in/21.34 m spread and 33ft 11 in/10.34 m swept; length 73 ft 6 in/22.40 m; height 17 ft 1⅜ in/5.22 m
POWERPLANT: two 20,350-lb/9,231-kg afterburning thrust Pratt & Whitney TF30-PW-7 turbofans
PERFORMANCE: speed 1,650 mph/2,655 km/h; ceiling 60,000 ft/18,290 m; range 2,925+ miles/4,707+ km
ARMAMENT: up to 37,500 lb/17,010 kg of disposable stores carried in a small internal weapons bay and on six swivelling underwing hardpoints; a typical load is 42 750-lb/340-kg free-fall bombs, six free-fall thermonuclear bombs or six SRAM nuclear defence-suppression missiles, plus a wide assortment of electronic countermeasures

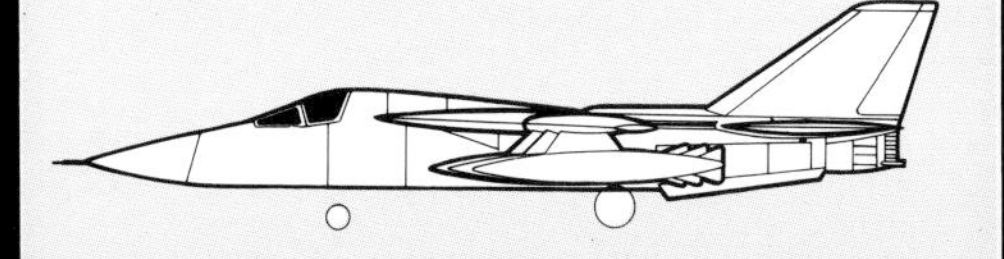

OPPOSITE, TOP The General Dynamics F-111's wing-sweep capability.

OPPOSITE, MIDDLE An FB-111A medium-range strategic bomber in cruise configuration with its bomb bay doors open to reveal two AGM-69A SRAM defence-suppression missiles.

LEFT The complexity of modern aerodynamics (both configuration and controls) is readily apparent in this illustration of an F-111D.

OPPOSITE, BOTTOM F-111s generally cruise with their wings unswept to maximize range, penetrating enemy airspace in the swept configuration for maximum speed and comfort of ride at low level.

BELOW End of day, and an F-111 taxies in with the canopies to its side-by-side seats open.

ABOVE From nose to tail the basically cylindrical fuselage of the Lockheed F-104G Starfighter accommodates electronics, the cockpit, fuel and the mighty afterburning turbojet.

ABOVE RIGHT Any view from below emphasizes the tiny and unswept wings of the F-104G and their Starfighter variants.

BELOW RIGHT Any photograph of a Starfighter showing the fuselage and wings explains why the type was dubbed the 'manned missile'.

continuities at the junctions. The MAW, on the other hand, provides smooth flexibility and the possibility of far more exact profile tailoring to suit operating conditions. Boeing's first move toward this concept was the leading-edge slats on its Model 747 airliner: these have flexible glassfibre skins, so they can be deployed in a smooth curve without discontinuity.

The MAW takes the process to its logical conclusion by enabling the whole wing to be flexed in profile hundreds of times per second. The structure is based on a rigid box attached to the wing-pivoting mechanism; to this are attached completely flexible leading- and trailing-edge sections with laminated glassfibre skins and a complex arrangement of internal actuators controlled by the aircraft's fly-by-wire system. The tactical advantages of such a wing are obvious, and even the prototype installation should allow range to be increased by up to 35% and provide a 25% improvement in sustained turn rate. In the longer term the MAW clearly offers both military and civil aircraft enormous advantages in performance and operating costs.

Until quite recent years, the emphasis in high-performance wing design had been on reducing thickness-chord ratio – that is the ratio of a wing's thickness to its breadth from leading to trailing edge – in order to reduce profile drag. Thin wings were common, especially on advanced military aircraft, though their use often meant that fuel and the main landing gear units had to be accommodated in

LOCKHEED F-104G STARFIGHTER

TYPE: *single-seat multi-role and interdiction fighter*
WEIGHTS: *empty 14,900 lb/6,758 kg; maximum take-off 28,779 lb/13,054 kg*
DIMENSIONS: *span 21 ft 11 in/6.68 m; length 54 ft 9 in/16.69 m; height 13 ft 6 in/4.11 m; wing area 196.1 sq ft/18.22 m^2*
POWERPLANT: *one 15,800-lb/7,167-kg afterburning thrust General Electric J79-GE-11A or MAN/Turbo-Union J79-MTU-J1K turbojet*
PERFORMANCE: *speed 1,450 mph/2,333 km/h; ceiling 58,000 ft/17,680 m; range 1,550 miles/ 2,495 km*
ARMAMENT: *one 20-mm Vulcan multi-barrel cannon and up to 4,310 lb/1,955 kg of disposable stores (including two Sidewinder short-range air-to-air missiles on the wingtip launcher rails and a wide assortment of conventional or – occasionally – nuclear free-fall and powered stores, both guided and unguided) on one underfuselage and four underwing hardpoints*

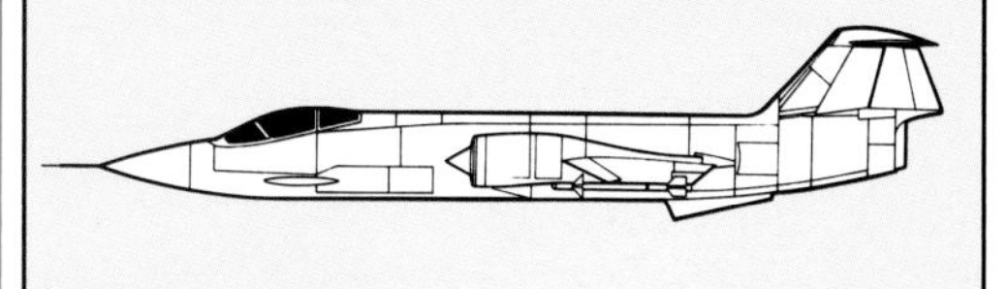

U.S. AIR FORCE

ABOVE The North American F-100 Super Sabre was a prodigious aerodynamic and structural achievement.

RIGHT Designed as an interceptor, the F-100 soon matured as a real workhorse of a fighter.

BELOW As it turns off onto a taxiway, this F-100 still streams its braking parachute.

the fuselage rather than in the wings. There was no problem with an aircraft like the Lockheed F-104 Starfighter, which has only a small and essentially unswept wing. However, such types as the North American F-100 Super Sabre and the contemporary Mikoyan MiG-19 were designed for greater agility and accordingly had bigger, highly swept wings which had low wave drag but involved the problem of aeroelastic distortion (that is, wing twist induced by reaction to aileron movement).

The designers of the F-100 skated round the problem by locating the ailerons inboard and omitting flaps, while the MiG-19 is a marvel of aircraft structure with ailerons and flaps. Other designers opted to fill in the area behind the swept trailing edge to create a delta wing of greater strength and internal volume, and one whose increased chord allowed greater depth for the same thickness/chord ratio. The classic example of this philosophy is the Mirage III, which has excellent straight-line performance at high altitude, but in sustained manoeuvres loses energy extremely rapidly as the large area of the wing is presented to the airflow.

LEFT Its large wing gives the Dassault Mirage III good performance at high altitude, but severely hampers its utility at low level.

BELOW LEFT The clean lines and massive delta wing of a Mirage IIIC interceptor; it carries two drop tanks and a single R530 air-to-air missile.

When flying at low levels, aircraft with delta or very slender wings exhibit a high response to gusts – local up- and down-currents that alter the wing's apparent angle of incidence. The resulting bumpiness in flight seriously curtails the life of the airframe and, more immediately, rapidly erodes its crew's ability to concentrate. Gust response is experienced by the crew as jerking vertical movements. The Mirage III suffers about 70 such 0.5-g bumps per minute, seven times the tolerable level, which can be achieved only by low-aspect-ratio aircraft such as the F-111 and Panavia Tornado. Yet the ability to fly for prolonged periods at low and very low levels is essential for modern aircraft faced with the task of penetrating hostile airspace.

The MAW not only has good gust-response characteristics, it is also ideally suited to the latest development in aerofoil section, namely the supercritical wing section. The supercritical wing is characterized by a fairly bluff leading edge, a bulged underside, a flat top and a down-curved trailing edge. The aerodynamics of such a section are complex, but they delay

LEFT The fairing under the rear fuselage of the Mirage III was designed to accept an SEPR liquid-propellant booster rocket, but the importance of high-altitude operations declined so rapidly in the late 1950s that this provision has seldom been employed.

RIGHT Though unsuccessful at the time, the Northrop YB-49 is now seen as a truly prophetic design – it is widely accepted that the new Northrop B-2 strategic bomber will have many conceptual similarities to the B-49 in layout.

maximum acceleration and shock formation, allowing the designer to create a wing with less sweep and a thicker section, which in turn provides extra volume and reduces structure weight. Supercritical wing sections are now used almost universally on military and civil aircraft designed for high subsonic and supersonic performance.

VARIABLE GEOMETRY

The MAW is also well suited to a variable-geometry layout. Variable geometry was pioneered in Germany during World War II, and further developed on an experimental basis in the USA during the late 1940s and early 1950s, but reached a practical level only with the F-111, which first flew in 1964.

It has been employed subsequently on several long-range combat aircraft, most notably the Tornado interdiction aircraft, the Rockwell B-1 penetration bomber, the Grumman F-14 Tomcat fleet-defence fighter and most new Soviet strike aircraft, among them the MiG-23/27 Flogger, Sukhoi Su-24 Fencer and Tupolev Blackjack.

NORTHROP YB-49

During development of the Northrop B-2 it was generally assumed that the US Air Force's new strategic stealth bomber would be a flying wing, and this would accord fully with the company's long association with this type of flying machine. The advantages of the flying wing over conventional aircraft stem from the enormous reductions in weight and drag resulting from the elimination of the fuselage and empennage (tail unit) and include reduced costs and much improved performance.

Jack Northrop had been an advocate of flying wings in the 1930s, and during World War II persevered with the ultimately unsuccessful XP-56 pusher fighter, which retained a vestigial fuselage. Greater success attended Northrop's larger flying wings, first of which was the XB-35 prototype strategic bomber. The XB-35 was exceptionally clean in line, being a true flying wing with tricycle landing gear; it had a range of 2,500 miles/4,023 km with a bombload of 20,000 lb/9,072 kg on the power of four 3,250-hp/2,424-kW Pratt & Whitney Wasp Major radials driving pusher propellers.

From the XB-35 Northrop developed the YB-49 jet bomber: powered by eight 4,000-lb/1,814-kg thrust Allison J35 turbojets, the YB-49 turned the scales at a maximum 216,600 lb/98,250 kg and reached a top speed of 520 mph/837 km/h, compared with the XB-35's 393 mph/632 km/h. Range was an excellent 2,800 miles/4,506 km with a 10,000-lb/4,536-kg bombload; an alternative bombload of 37,400 lb/16,965 kg could be carried over shorter ranges. However, there were problems with the design, and ultimately the production contract was cancelled.

(pictured opposite)

ABOVE The Rockwell B-1A was developed at a time when aviation technologies were evolving rapidly, and considering its cancellation with hindsight, it is possible to see that President Carter's much-vilified decision was in all probability the right one.

RIGHT The variable-geometry B-1A offered superb capabilities, but only at altitudes at which it would have been vulnerable to a whole host of sophisticated weapons.

ABOVE The value of the B-1A lies in the fact that it has paved the way for the B-1B operational bomber. This is still plagued with structural and avionics problems, but offers the real possibility of deep penetration into enemy airspace at high subsonic speed and very low levels: and by comparison with the B-1A the type has considerably lower radar signature to make its detection that much more difficult.

SUKHOI Su-7BMK FITTER-A

TYPE: *single-seat close-support fighter*
WEIGHTS: *empty 19,004 lb/8,620 kg; maximum take-off 29,762 lb/13,500 kg*
DIMENSIONS: *span 29 ft 3½ in/8.93 m; length 57 ft 0 in/17.37 m including nose probe; height 15 ft 0 in/4.57 m; wing area 339.1 sq ft/ 31.50 m²*
POWERPLANT: *one 22,046-lb/10,000-kg afterburning thrust Lyul'ka AL-7F-1 turbojet*
PERFORMANCE: *speed 1,055 mph/1,700 km/h; ceiling 49,700 ft/15,150 m; range 430 miles/ 690 km*
ARMAMENT: *two 30-mm NR-30 cannon and up to 5,511 lb/2,500 kg of disposable stores (including a single tactical nuclear weapon, rocket pods, cannon pods, air-to-surface missiles and a wide assortment of free-fall conventional or cluster bombs) on two underfuselage and four underwing hardpoints*

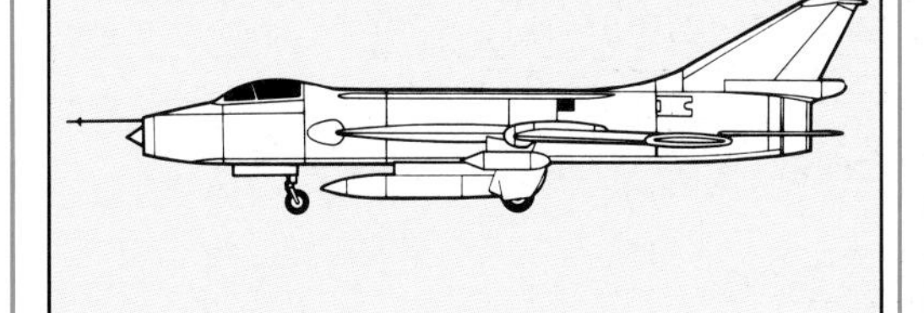

SUKHOI Su-17 FITTER-C

TYPE: *single-seat close-support fighter with variable-geometry wings*
WEIGHTS: *empty 24,030 lb/10,900 kg; maximum take-off 39,020 lb/17,700 kg*
DIMENSIONS: *span 45 ft 11¼ in/14.0 m spread and 34 ft 9½ in/10.60 m swept; length 63 ft 0 in/19.20 m including nose probe; height 17 ft 6⅝ in/5.35 m; wing area 431.65 sq ft/40.10 m² spread and 400.4 sq ft/37.20 m² swept*
POWERPLANT: *one 24,691-lb/11,200-kg afterburning thrust Lyul'ka AL-21F-3 turbojet*
PERFORMANCE: *speed 1,432 mph/2,305 km/h; ceiling 59,055 ft/18,000 m; range 780 miles/ 1,255 km*
ARMAMENT: *two 30-mm NR-30 cannon and up to 8,818 lb/4,000 kg of disposable stores (including a single tactical nuclear weapon, rocket pods, cannon pods, air-to-surface missiles and a wide assortment of free-fall conventional or cluster bombs) on four underfuselage and four underwing hardpoints*

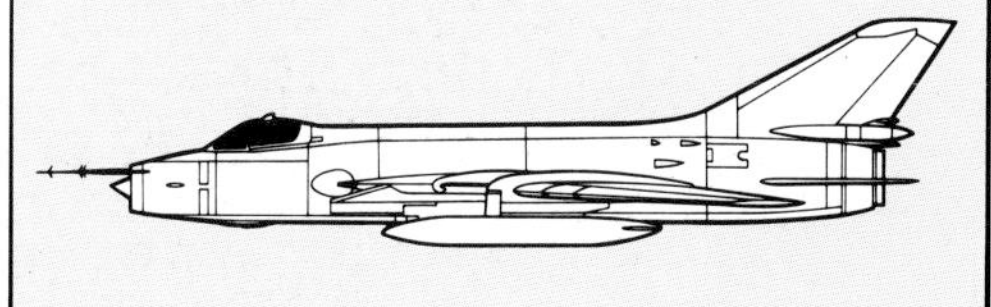

A typical variable-geometry, or swing-wing planform, such as that of the F-111, is based on a comparatively wide centre section accommodating the wing pivots and sweep actuators behind sharply swept wing gloves. Outboard of the pivots are the moving wing panels, which can be swept from a fully-forward 16° sweep angle to a maximum of 72.5°. In the forward positon the wings provide good low-speed lift and load-carrying capability, offering long range and the possibility of operations from short runways. When fully swept they provide low wave drag and a planform optimized for supersonic flight, as well as low gust response in high-speed low-level flight. The Western aircraft with the best gust response characteristics are the Tornado (bearable), the F-111 and B-1 (bearable for short periods) and the F-14 (on the verge of unacceptable); there is no reason to doubt that Soviet aircraft of this type enjoy the same benefits.

STOL CAPABILITY

Swing wings can also accommodate high-lift devices. Such devices are very important in increasing wing area and lift in low-speed flight, thereby providing short take-off and landing (STOL) capability.

This is an increasingly important feature of modern combat aircraft, though Western

ABOVE LEFT The Sukhoi Su-24 is the Soviet equivalent of the American F-111, like the US aircraft offering the possibility of long-range interdiction with its good avionics and variable-geometry layout.

ABOVE The MiG-23 is another variable-geometry Soviet aeroplane and a versatile multi-role fighter.

BELOW The MiG-23's ventral tail folds sideways on the ground to avoid contact with the runway.

ABOVE The use of a shoulder-set wing means that the MiG-23's main landing gear units are attached to the lower fuselage, with a somewhat complex arrangement to provide a wide enough track to avoid instability problems on the runway.

MIKOYAN MiG-23MF FLOGGER-B

TYPE: *single-seat air-combat and multi-role fighter with variable-geometry wings*
WEIGHTS: *empty 24,250 lb/11,000 kg; maximum take-off 41,667 lb/18,900 kg*
DIMENSIONS: *span 46 ft 9 in/14.25 m spread and 26 ft 9½ in/8.17 m swept; length 59 ft 6½ in/18.15 m including the nose probe; height 14 ft 4 in/4.35 m; wing area 293.4 sq ft/ 27.26 m^2*
POWERPLANT: *one 25,353-lb/11,500-kg afterburning thrust Tumanskii R-29 turbojet with variable-geometry inlets and variable nozzle*
PERFORMANCE: *speed 1,522 mph/2,450 km/h; ceiling 60,040 ft/18,300 m; range 1,180 miles/ 1,900 km*
ARMAMENT: *one 23-mm GSh-23L twin-barrel cannon and up to 6,614 lb/3,000 kg of disposable stores (including six air-to-air missiles and an extremely wide assortment of disposable stores such as air-to-surface missiles, rocket pods and free-fall conventional or nuclear bombs) carried on one underfuselage, two under-trunk and two underwing hardpoints*

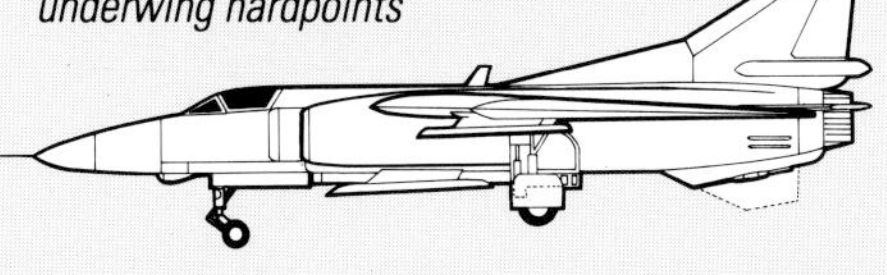

MIKOYAN MiG-27 FLOGGER-D

TYPE: *single-seat close-support and ground-attack fighter with variable-geometry wings*
WEIGHTS: *empty 23,778 lb/10,790 kg; maximum take-off 44,313 lb/20,100 kg*
DIMENSIONS: *span 46 ft 9 in/14.25 m spread and 26ft 9¾ in/8.17 m swept; length 52 ft 5⅞ in/16.0 m; height 14 ft 9.2 in/4.50 m; wing area 293.33 sq ft/27.25 m^2 spread*
POWERPLANT: *one 25,353-lb/11,500-kg afterburning thrust Tumanskii R-29 turbojet with plain inlets and two-position nozzle*
PERFORMANCE: *speed 1,123 mph/1,807 km/h; ceiling 52,495 ft/16,000 m; range 480 miles/ 780 km*
ARMAMENT: *up to 8,818 lb/4,000 kg of disposable stores (typically a wide assortment of air-to-surface weapons of the unguided free-fall conventional or nuclear and guided powered types, plus rocket pods) carried on three underfuselage, two under-trunk and two underwing hardpoints*

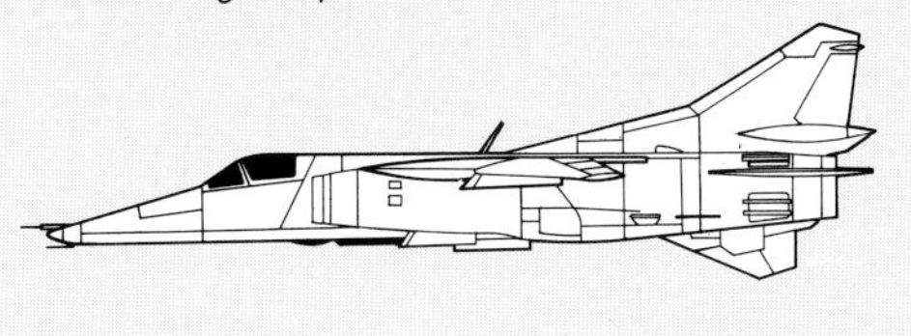

SAAB 37 VIGGEN OFF-RUNWAY DISPERSAL

Just about every air base in western Europe is a likely target for Soviet attack in the event of war: it is probable that within hours if not minutes of the outbreak of hostilities every substantial runway would be destroyed by a nuclear attack or severely cratered by conventional weapons. Western European combat aircraft, together with American and Canadian machines based on the eastern side of the Atlantic, would be unable to operate even if they had survived the attack in their hardened aircraft shelters. Cratered runways can be repaired, but they can be re-cratered just as easily, and in such circumstances the only Western aircraft still able to operate would be the British Aerospace Harrier, with its short take-off and vertical landing capability, and the Swedish Saab 37 Viggen.

The Viggen is the aircraft component of the Swedes' System 37 defence scheme, and was designed to operate without loss of capability from dispersed sites, comprising 545-yard/500-m lengths of straight road in any part of the country. System 37 also involves large numbers of vehicles to ensure that maintenance crews, along with fuel and ordnance, can be ferried to these dispersed sites as required in times of crisis. To ease the maintenance load at such sites the Viggen has been designed for maximum reliability and maintainability, with a light airframe stressed to 12 g and ground-level access to most major systems in the airframe.

The Viggen's canard configuration is admirably suited to steep, slow arrivals and departures at the dispersed sites, and its landing gear is built by Motala Verstad to meet extremely exacting requirements for no-flare landing at a sink rate of 16.4 ft/5 m per second. Each main unit features a long-stroke oleo, shortened during retraction to require less internal volume, and a tandem arrangement of two wheels pressurized to 215 lb/sq in 15.12 kg/cm^2 to help absorb the considerable landing shock. As soon as the twin nosewheels hit the ground, the compression of the nose unit oleo activates the switch that operates the engine thrust-reverser system. (The Viggen's powerplant is a Volvo Flygmotor RM8A afterburning turbofan, derived from the civil Pratt & Whitney JT8D for maximum reliability.) As soon as the switch is operated, the engine's exhaust is deflected forward through three annular slots in the rear fuselage, complementing the brakes to halt the Viggen safely and quickly.

countries have been slow to accept a fact that has been evident to the Soviets and Swedes for decades. Almost all Western military aircraft depend on expensive air bases with extensive hangarage and vast concrete runways which are conspicuously vulnerable to conventional weapons, leaving aside the effect of a nuclear attack.

It is estimated that all NATO's airfields have been targeted by at least two Soviet nuclear-armed missiles, raising the very real possibility that once a nuclear exchange had been initiated all NATO air bases would either cease to exist or, at best, be severely damaged. Conventional aircraft cannot operate from short stretches of damaged runway; so there is every likelihood that the vast majority of NATO's tactical aircraft would be rendered impotent, even if their hardened aircraft shelters protected them from the initial strike.

The Swedes have opted for STOL aircraft, in the shape of the Saab 35 Draken and Saab 37 Viggen, which can operate from dispersed sites such as straight stretches of country road. The Soviet answer is a generation of tactical aircraft whose high-lift devices and landing gear are suitable for rough field operations. Some

SAAB AJ37 VIGGEN

TYPE: *single-seat all-weather attack aircraft*
WEIGHTS: *empty 26,015 lb/11,800 kg; maximum take-off 45,194 lb/20,500 kg*
DIMENSIONS: *span 34 ft 9¼ in/10.60 m for main delta wing and 17 ft 10½ in/5.45 m for canard foreplane; length 53 ft 5¾ in/16.30 m; height 19 ft 0¼ in/5.80 m; wing area 495.1 sq ft/ 46 m^2 for main delta wing and 66.74 sq ft/ 6.20 m^2 for canard foreplane*
POWERPLANT: *one 26,015-lb/11,800-kg afterburning thrust Volvo Flygmotor RM8A turbofan (licence-built Pratt & Whitney JT8D-22 with locally designed afterburner and thrust reverser)*
PERFORMANCE: *speed 1,320 mph/2,125 km/h; ceiling 49,870 ft/15,200 m; range 1,243+ miles/2,000+ km*
ARMAMENT: *up to 13,228 lb/6,000 kg of disposable stores (including air-to-air missiles, air-to-surface missiles, cannon pods, rocket pods and various types of free-fall bomb) on three underfuselage and four underwing hardpoints*

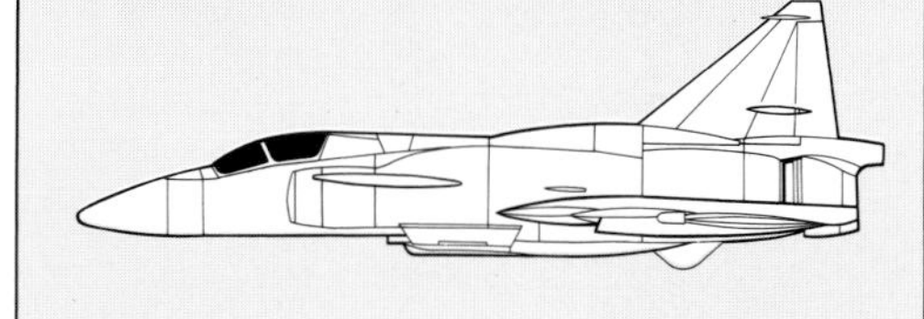

NATO countries are finally coming to grips with the problem too, and the Americans are currently investigating a derivative of the F-15 Eagle whose two-dimensional thrust-vectoring system provides the ability to operate from 500-yard/455-m lengths of undamaged runway.

HIGH-LIFT TORNADO

The NATO type best able to deal with damaged runways is the swing-wing Tornado, whose primary control surfaces are slab tailplane halves that operate in unison for pitch control and differentially for roll control and a powerful rudder for yaw control. The wings are left free

ABOVE For its time the Saab 35 Draken was a highly innovative design, the double-delta wing offering great lift and large fuel volume without a massive drag penalty. This is the J 35E reconnaissance version.

RIGHT The Saab 37 Viggen is another innovative design from Sweden, this time using a canard configuration and thrust-reversing turbofan for exceptional short-field performance.

LEFT The SF 37 is the overland reconnaissance version of the Viggen, external reconnaissance equipment being used to provide the maximum flexibility of sensor carriage.

RIGHT The Panavia Tornado GR Mk 1 is in its element at high speed and very low level.

BELOW The real skill in designing modern combat aircraft lies not just in producing a machine capable of meeting its specification, but of meeting the specification in as compact an airframe as possible. The Tornado is a superb example of such skill, yet has considerable growth potential.

FAR RIGHT Below the nose of this Tornado GR Mk 1 is the fairing over the laser ranger and marked-target seeker.

BELOW RIGHT This Tornado GR Mk 1 carries two JP233 airfield attack dispensers, each carrying runway-cratering submunitions and area-denial minelets to hamper the repair crews.

PANAVIA TORNADO IDS

TYPE: *two-seat all-weather interdiction and multi-role fighter with variable-geometry wings*
WEIGHTS: *empty 31,063 lb/14,090 kg; maximum take-off 60,000 lb/27,215 kg*
DIMENSIONS: *span 45 ft 7½ in/13.91 m spread and 28 ft 2½ in/8.60 m swept; length 54 ft 10¼ in/16.72 m; height 28 ft 2½ in/5.95 m; wing area about 269 sq ft/25.0 m^2*
POWERPLANT: *two 16,800-lb/7,620-kg afterburning thrust Turbo-Union RB 199-34R Mk 103 turbofans*
PERFORMANCE: *speed 1,453+ mph/ 2,337+ km/h; ceiling 49,210+ ft/15,000 m; range 1,727+ miles/2,780+ km*
ARMAMENT: *two 27-mm Mauser BK27 cannon and up to 19,840 lb/9,000 kg of disposable stores (including Sidewinder short-range air-to-air missiles, Kormoran or Sea Eagle anti-ship missiles, Maverick air-to-surface missiles, guided glide bombs, unguided free-fall conventional or cluster bombs, dispenser weapons, rocket pods and a wide assortment of electronic countermeasures pods) on three underfuselage and four swivelling underwing hardpoints*

for the high-lift devices that are claimed to give the Tornado the best lifting qualities of all supersonic aircraft with its wings in the 25° minimum-sweep position.

The leading edge of each Tornado wing is occupied over its full span by three-section leading-edge slats. The trailing edge has full-span, four-section, double-slotted flaps complemented by two-section lift-dumpers which also operate as spoilers to supplement the tailerons in roll control. And the high-lift system is completed by a Krüger flap on each 60° swept wing glove. Widely used on civil aircraft, the Krüger flap is a leading-edge device forming part of the undersurface of the leading edge, being hinged to swing down and forward to create a bluff leading edge suitable for low airspeeds on a high-speed wing. The whole system is enhanced by a powerful airbrake on each side of the vertical tail, and by thrust reversers on the two afterburning turbofan engines.

Such high-lift devices are standard on civil and military transport aircraft, which may have triple-slotted flaps complemented by drooping ailerons. Lift is enhanced by the increasing lifting area of the extended flaps, which also control the airflow for optimum lift at low airspeeds, while drag is generated by the additional area and deflected position of the flaps in order to lower the aircraft safely onto the runway at the minimum possible airspeed. Safety is thereby increased, while runway requirements are minimized.

Some airliners are also provided with winglets, devices which are set to become more common on long-range military aircraft such as strategic transports and maritime patrol aircraft. The term is a misnomer, since winglets are actually upturned – or occasionally downturned – wingtip extensions designed to improve the cruise efficiency of the wing. They do so by reducing the tip vortex, consequently curtailing induced drag and recovering the energy otherwise lost to this cause, and by enhancing the circulation of air over the outer wing panels to generate additional lift.

ABOVE The classic configuration of high-lift devices on transport aircraft is revealed on this Boeing 727-200 airliner, which sports Krueger leading-edge flaps (inboard) and leading-edge slats (outboard) as well as triple-slotted trailing-edge flaps.

RIGHT A test model of the McDonnell Douglas C-17 long-range transport, currently under development for the US Air Force, reveals the configuration of the modern tactical transport, with a T-tail, massive fuselage accommodating the main landing gear units in blister fairings, and an uncluttered wing with winglets.

WINGLETS

Developed largely on the basis of research by Richard T. Whitcomb, one of the most prolific of modern aerodynamicists, the winglet is a highly effective way of boosting the cruise efficiency of a wing. The winglet comprises an extension of the main wing at the tip; turned either up or down, it serves the double purpose of reducing the wingtip vortex, thereby recovering the energy that would otherwise be lost in the vortex, and improving the circulation – and therefore the lift – of the outer wing panel. The wingtip vortex is a high-energy rotational movement of air which streams back, out and generally down from the tip, producing considerable drag: generally fitted over the rear half of the tip chord and angled outward slightly, the winglet serves both to reduce and to control this vortex.

The winglet also helps prevent spanwise separation of the airflow over the upper surface of the wing – a similar purpose is served by the prominent fences on the wings of many Soviet aircraft – and reduces the leakage of higher-pressure air from the under surface of the wing round the tip to the lower-pressure air flowing over the upper surface. As this pressure differential is the feature that creates lift, its maximization by the winglet, allied to other devices, allows overall area and structural weight to be reduced.

VTOL AND STOVL

The problem of runway vulnerability can be avoided entirely by vectoring – that is, pointing – engine thrust to permit vertical take-off and landing (VTOL) or, in more practical operational terms, short take-off and vertical landing (STOVL). The classic example of this type of aircraft is the British Aerospace Harrier, which is now in service in its much revised Anglo-American form as the McDonnell Douglas/British Aerospace Harrier II.

The rationale for such an aeroplane is unimpeachable: if the total thrust of the engine exceeds the take-off weight of the aeroplane, and if this thrust can be vectored downwards to provide direct lift, then the aeroplane can rise straight up from the ground without using a runway at all. VTOL aircraft can be based anywhere in times of crisis, virtually eliminating their vulnerability to pre-emptive destruction and adding enormously to their tactical versatility so long as fuel, ordnance and other supplies can be provided at their operating sites.

Inevitably, there is a price to be paid for the tactical advantages offered by VTOL. The most obvious drawback is that for vertical take-off the thrust/weight ratio must exceed unity, and this generally means that the Harrier can only lift off vertically with a reduced weapon load or a reduced fuel load. In theory this disadvantage can be overcome by the use of in-flight refuelling: the Harrier could take off with maximum weapons but only minimum fuel, and after translation from engine- to wing-borne flight could rendezvous with a tanker aircraft to fill its tanks. Such a scheme would tie the dispersed-site Harrier to runway-based tankers, however. Instead, the answer lies in a compromise between thrust- and wing-borne flight to allow take-off with useful weapon and fuel loads.

Operational practice involves a very short take-off run: with the engine nozzles at 0°, pointing straight aft for maximum thrust, the pilot accelerates for a short distance then pulls the thrust lever back to the 90° position. All the thrust is instantly converted into a vertical component that jumps the aeroplane into the air, where the pilot can ease the lever back towards the 0° position and so translate rapidly into wing-borne flight. Landing vertically presents no problem, for the weapon load will

ABOVE The basic concept of the British Aerospace Harrier has been taken one step further in the McDonnell Douglas AV-8B Harrier II, an advanced development most notable for its superior cockpit, larger single-piece wing of composite construction and supercritical section, and considerably enhanced lift-improvement devices. Two side strakes serve as the mountings for the 25-mm cannon and its ammunition supply.

BRITISH AEROSPACE HARRIER GR Mk 3

TYPE: *single-seat close-support and reconnaissance aircraft with STOVL (short take-off and vertical landing) capability*
WEIGHTS: *empty 13,535 lb/6,139 kg; maximum take-off 25,200+ lb/11,431+ kg*
DIMENSIONS: *span 25 ft 3 in/7.70 m or 29 ft 8 in/9.04 m with low-drag bolt-on ferry tips; length 46 ft 10 in/14.27 m; height 11 ft 4 in/3.45 m; wing area 201.1 sq ft/18.68 m^2 or 216.0 sq ft/20.07 m^2 with bolt-on ferry tips*
POWERPLANT: *one 21,500-lb/9,752-kg thrust Rolls-Royce Pegasus II Mk 103 non-afterburning turbofan with four vectoring nozzles, the forward pair for cold gas from the fan and the aft pair for hot gas from the core*
PERFORMANCE: *speed 737+ mph/1,186 km/h; ceiling 50,000+ ft/15,240+ m; range 828 miles/1,316 km*
ARMAMENT: *two 30-mm Aden cannon and up to 8,000 lb/3,269 kg of disposable stores (including Sidewinder short-range air-to-air missiles, Martel air-to-surface missiles, rocket pods, Paveway laser-guided glide bombs and a wide assortment of free-fall conventional or cluster bombs) on one underfuselage and four underwing hardpoints*

BRITISH AEROSPACE SEA HARRIER FRS Mk 1

TYPE: *single-seat carrier-borne fighter, reconnaissance and strike aircraft with STOVL (short take-off and vertical landing) capability*
WEIGHTS: *empty 13,100 lb/5,942 kg; maximum take-off 26,190 lb/11,880 kg*
DIMENSIONS: *span 25 ft 3 in/7.70 m; length 47 ft 7 in/14.50 m; height 12 ft 2 in/3.71 m; wing area 201.1 sq ft/18.68 m^2*
POWERPLANT: *one 21,500-lb/9,752-kg thrust Rolls-Royce Pegasus II Mk 104 non-afterburning turbofan with four vectoring nozzles, the forward pair for cold gas from the fan and the aft pair for hot gas from the core*
PERFORMANCE: *speed 735 mph/1,183 km/h; ceiling 50,000+ ft/15,240+ m; range 920 miles/1,480 km*
ARMAMENT: *two 30-mm Aden cannon and up to 8,000 lb/3,629 kg of disposable stores (including Sidewinder short-range air-to-air missiles, Martel air-to-surface missiles, Harpoon or Sea Eagle anti-ship missiles, rocket pods, Paveway laser-guided glide bombs, the WE-177 tactical nuclear bomb and a wide assortment of free-fall conventional or cluster bombs) on one underfuselage and four underwing hardpoints*

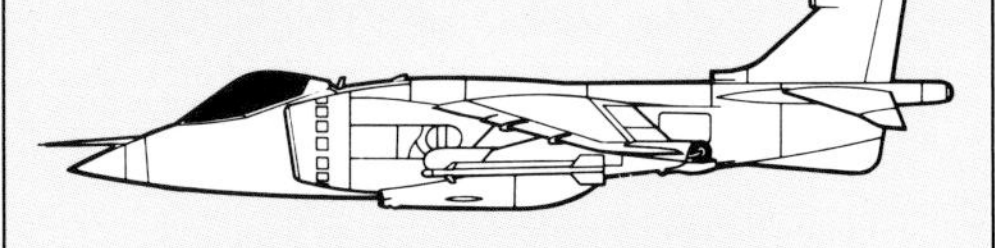

ABOVE The AV–8B in classic operating locale: any small space that can be cleared close to the front line.

RIGHT The British version of the AV–8B is the Harrier GR Mk 5, which has a number of detail modifications compared with the American model and which, like the AV-8B, can ultimately be fitted with radar for an all-weather attack capability.

By removing the need for a fixed operating base with its vulnerable runways, the designers of the STOVL Harrier created a new air weapon. The story of Harrier operations is encapsulated in this sequence of photographs:

LEFT The Harrier GR Mk 3 operates from concealed sites such as woodland clearings close to a road or track that can be used by the squadron's fuel, ordnance and other logistical vehicles.

BELOW LEFT The Harrier can operate from its clearing in the vertical take-off mode, but is able to uplift a heavier warload after a short take-off from a length of road.

BELOW RIGHT The Harrier can carry a substantial load, but has the performance and agility to cope with dedicated fighters mustered by the enemy.

RIGHT With its mission completed the Harrier lands vertically and immediately taxies off the strip into the concealment of the woods.

ABOVE The Harrier's simple airfield operations mean that the type can do more than adequately with small plain flaps, leaving the wing uncluttered by the normal plethora of high-lift devices, thus designed and built for flight.

LEFT Viewed from below, the Harrier reveals the massive fuselage required for the turbofan, its large inlets and the four-poster deflection system for the cold and hot gases via the front and rear nozzles respectively.

RIGHT An overhead shot of the AV-8B construction line highlights the one-piece wing, based on a composite-structure torque box accommodating a large volume of fuel.

NO SMOKING
ABOVE FLOOR LEVEL

have been expended and the fuel load greatly reduced, restoring the thrust/weight ratio to greater than unity.

Various expedients have been adopted in the designs of the Harrier, Sea Harrier and Harrier II to increase vertical lift. As much as possible of the gas vectored down from the engine is trapped, and control at nil or very low air-speeds is provided by reaction nozzles using cool air tapped from the engine compressor. In general, though, the use of a thrust-vectoring powerplant allows the flying surfaces to be much simpler than those of a comparable fixed-thrust type. Thrust-vectoring also makes for enormous agility in the air, and the Harrier variants have proved in many air-combat manoeuvres that they can use the technique known as VIFFing (vectoring in forward flight) to master even dedicated air-combat fighters like the F-16.

In combat, VIFFing is used for direct lift control comparable to that of experimental machines such as the AFTI/F-16. If the pilot of a conventional fighter finds an opponent on his tail, he may try to shake it off with a series of tight turns, but air-combat fighters generally have the agility to follow and even to close in such manoeuvres. In a comparable situation,

ISRAEL AIRCRAFT INDUSTRIES Kfir-C2

TYPE: *single-seat interceptor and ground attack aircraft*
WEIGHTS: *empty 16,060 lb/7,285 kg; maximum take-off 35,714 lb/16,200 kg*
DIMENSIONS: *span 26 ft 11½ in/8.22 m for the main delta wing and 12 ft 3 in/3.73 m for canard foreplane; length 51 ft 4½ in/15.65 m including nose probe; height 14 ft 11¼ in/ 4.55 m; wing area 374.6 sq ft/34.80 m^2 for main delta wing and 17.87 sq ft/1.66 m^2 for canard foreplane*
POWERPLANT: *one 17,900-lb/8,119-kg afterburning thrust General Electric J79-GE-J1E turbojet*
PERFORMANCE: *speed 1,516+ mph/ 2,440+ km/h; ceiling 58,000+ ft/17,680 m; range 428 miles/690 km*
ARMAMENT: *two 30-mm DEFA 552/553 cannon and up to 12,731 lb/5,775 kg of disposable stores (including Shafrir 2 or Python 3 short-range air-to-air missiles, Luz-1 or Maverick air-to-surface missiles, Shrike anti-radiation missiles, rocket pods, guided glide bombs and a wide assortment of free-fall conventional or cluster bombs) on five underfuselage and four underwing hardpoints*

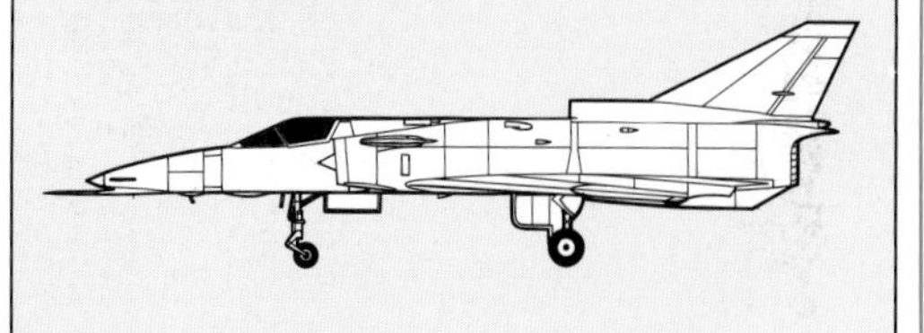

ABOVE Comparison with earlier illustrations of the F-16 confirms the radically different appearance of the AFTI/F-16, the result of the addition of the canted control surfaces under the inlet. Less obvious is the larger dorsal strake accommodating this research aeroplane's mass of additional avionics.

ABOVE RIGHT By adding canard foreplanes to its Kfir (itself a development of the Mirage III/5 series), IAI has produced the much enhanced Kfir-C2 with greater combat agility and far superior field performance.

MIDDLE RIGHT The Kfir-TC2 is the two-seat version of the Kfir-C2, the larger nose and other volumes being filled with additional electronics for the type's primary role of electronic warfare under the guidance of the officer in the rear seat.

BELOW RIGHT France's refusal to sell Israel the clear-weather Mirage 5 prompted full-scale development of the Kfir at a rapid pace; but the baseline Mirage 5 then went on to secure substantial orders from many nations. This example is in the markings of Venezuela.

the Harrier pilot can roll into the turn and select thrust 90° down: all the engine's thrust is then directed straight into the direction of the turn, reducing its radius to a degree unmatchable by a conventional fighter. And as the thrust-vectoring system allows a maximum 110° deflection (20° forward of straight down), the system can also be used to brake the Harrier more rapidly than a conventional fighter's airbrakes can manage.

The combination of a compact and inherently agile airframe with a thrust-vectoring system gives the STOVL aircraft an unprecedented level of agility. It is truly remarkable that the combination has not yet been adopted for dedicated air-combat fighters, which do not need to carry heavy offensive loads and could therefore afford to take off vertically.

CANARD FOREPLANES

The canard configuration, involving a tail-mounted delta wing and nose-mounted delta foreplanes, as used in the vast B-70 strategic bomber, is becoming standard for smaller fixed-wing combat aircraft. There are two good aerodynamic reasons for its application to such types as the Swedish Saab 37 Viggen and the Israel Aircraft Industries Kfir. Firstly, the nose of a normal tailless delta aircraft can be raised only by the application of up-elevon, which produces a down-load at the tail that must be subtracted from overall wing lift in any lift equation. This translates into a longer take-off run – since the landing gear is being

Rafale

pushed down onto the runway, the aeroplane has to accelerate to a higher speed to compensate for what is effectively a smaller wing – and an increase in the radius of any turn, since the tail is being pushed away from the origin of the turn radius. However, with a canard configuration the take-off run is shortened, since the elevons on both the tail and nose deltas are deflected downwards, creating up-loads that must be added to the overall wing/canard lift; and the turning radius is reduced as the foreplane pulls the nose in towards the origin of the turn radius.

Secondly, the canards create powerful vortices that stream back and out over the upper surfaces of the wings, re-energizing the sluggish boundary-layer vortices of those surfaces and delaying the breakaway of the airflow over their all-important outer panels. This is particularly important at high angles of attack, and a similar though less effective result can be secured in aircraft of a more conventional configuration by the use of leading-edge vortex generators (small bladelike projections, otherwise called turbulators), dogtoothed leading edges, which also reduce the thickness/chord ratio of the outer panels for useful advantages at transonic speeds, and LERXes or fuselage chines.

The canard has now become extremely fashionable in combat aircraft circles. The configuration has been adopted for the Viggen's successor, the JAS 39 Gripen; it will be used for Europe's two next-generation fighters, the Dassault-Breguet Rafale and the multi-national European Fighter Aircraft; and it may well be featured by the US Air Force's Advanced Tactical Fighter, as well as the Soviet aircraft.

TRANSONIC TAILS

A prominent characteristic of modern combat aircraft is their large horizontal and vertical tail surfaces. On supersonic aircraft the former are generally of the fully-powered slab type, in which the whole surface moves to ensure adequate pitch control during the severe trim changes that occur at transonic speeds. Such control problems were only appreciated in the early 1950s as jet aircraft began to approach the speed of sound barrier; they were accentuated by the accompanying reduction in wing area, especially of swept-wing aircraft. Many aircraft were lost to Dutch rolling, a lateral oscillation of the aeroplane with increasingly uncontrollable yaw and roll components. Larger tail surfaces did much to cope with the problem, and are particularly evident on Soviet aircraft: the MiG-25 has massive outward-canted vertical tail surfaces, while the MiG-23 is fitted with a ventral fin so large that it has to be folded sideways for take-off and landing.

DIRECT CONTROL MODES

Controllability is a key feature of modern aerodynamics, and lies at the heart of the CCV technology discussed above. But the standard controls of current aircraft can only rotate the aircraft around the relevant axis – longitudinal for roll, transverse for pitch and vertical for yaw – which means that it is some time before changes in one or more planes modify the aircraft's trajectory. This can waste a lot of time.

For example, if a pilot is about to attack a target with gunfire and finds that the crosswind has pushed his aircraft too far to one side, he has to roll in the right direction for what he estimates to be the correct distance, and then roll in the opposite direction to get back onto

OPPOSITE, BELOW Typical of the shape that will become standard in the 1990s is the Dassault-Breguet Rafale. This epitomizes the new generation of air-combat fighter with fly-by-wire controls, a close-coupled canard configuration, extensive use of composites, and advanced electronics of both offensive and defensive varieties.

OPPOSITE, ABOVE Another keynote of the Rafale's design is the elegance of its aerodynamic lines, which boost performance while helping to reduce radar cross-section.

ABOVE The USSR is also heavily committed to the new generation of blended aerodynamics, with aircraft such as this Mikoyan-Gurevich MiG-29, an air superiority and multi-role fighter possessing marked affinities with the McDonnell Douglas F-15 in design, but rather more like the same company's F/A-18 Hornet dual fighter and attack aircraft in size.

BELOW LEFT A view from below of the MiG-29 emphasizes the type's aerodynamic/structural blend, based on a wide fuselage/centre section with leading-edge root extensions, modestly swept wings, widely spaced underset engines and a powerful, all-moving tail.

the right trajectory. It would be much quicker if the pilot had additional control surfaces able to exert direct force normal to the trajectory: in the example quoted he could use his controls to exert a lateral force, shifting his aeroplane sideways without altering his heading until the target was right in his sight. Conversely, he could alter the lateral displacement of the nose without modifying his basic flight trajectory in order to keep the target in his sight while maintaining a course offset slightly to one side of it.

Similar techniques were pioneered in the form of direct lift control by airliners such as the Lockheed L-1011-500 TriStar, which use their spoilers to modify their glidepath, and have long been a feature of glider landings, where the lift is dumped to achieve a high rate of descent with the fuselage level. But the concept has been taken a stage further by the advanced fighter technology integration version of the Fighting Falcon.

The AFTI/F-16, has additional control surfaces in the form of two outward-canted oblique foreplanes located under the inlet duct and two outward-canted fixed ventral fins. The control surfaces are used in conjunction with a highly advanced digital flight-control system, and the aircraft has demonstrated unprecedented flight agility. In the pitch plane, for example, it can elevate its nose without gaining height and gain height without lifting its nose. Comparable modifications are attainable in the other two planes, and the most significant limitation so far encountered is the inability of the pilot to sustain accelerations of 2 g or more, normal to the line of flight, while remaining fixed in his seat and looking forward through the head-up display. Nevertheless, the AFTI/F-16 clearly points the way ahead for the aerodynamic control of high-performance aircraft.

CONFORMAL WEAPON CARRIAGE

Another radical development is the cranked-arrow F-16XL. The F-16's normal wing and tailplane have been replaced by a large cranked-arrow delta wing, which is essentially an ogival delta wing with extended tips: the wings curve out from a point under the cockpit at a sweep angle of about 50°, then angle back at 70° to a point as far outboard as the two structural beam fairings that project aft from the trailing edge, before continuing to the tip at the reduced sweep of 50°.

GENERAL DYNAMICS F-16/AFTI COCKPIT

Modified as part of the American Advanced Fighter Technology Integration (AFTI) programme, the General Dynamics F-16/AFTI flying testbed is immediately recognizable by the obliquely-canted canards under its inlet trunk and an enlarged dorsal spine housing additional electronics. The configuration had been pioneered with the YF-16/CCV (control-configured vehicle) adaptation of a pre-production Fighting Falcon, but the F-16/AFTI's triplex digital fly-by-wire flight-control system made possible fully decoupled (non-rotative) flight manoeuvres.

Flight trials confirmed that the system opened new ways to fly. In the vertical plane, for example, a conventional aircraft has to raise its nose to climb, whereas the F-16/AFTI can gain or lose height without altering its attitude, which could be useful during weapon delivery. Conversely, the F-16/AFTI can alter its angle of attack – the angle at which its wings meet the airflow – without rising or falling, a useful facility in air-to-air gunnery.

The cockpit of the F-16/AFTI is typical of the basic F-16, but has been modified to provide CCV monitoring. The most obvious modification is the special wide-angle head-up display developed in the UK by GEC Avionics. The standard F-16 HUD has a comparatively small field of view – 13.5° in azimuth and 9° in elevation – and was clearly unsuitable for the decoupled flight capability of the F-16/AFTI. The company therefore developed a more advanced HUD with 20° × 15° field of vision. The latter was used as the basis for the 30° × 20° HUD installed on F-16Cs and Ds, the holographic unit originally planned having proved impossible to produce.

LEFT The Lockheed TriStar is typical of modern airliner design so far as the powerplant is concerned; the two wing-mounted engines are suspended on projecting pylons where their mass helps to alleviate wing twist and bend. Such a location also allows easy maintenance of the engine and comparatively straightforward substitution of another engine (of either the same or a newer and probably larger type), and leaves the wing uncluttered to perform its primary aerodynamic functions.

BELOW Airliners such as the TriStar are models of careful design, commercially successful types blending safety and the last word in operating economics.

ABOVE Compared with the initial Tornado IDS variant, the Tornado ADV has a longer fuselage, and this pays handsome dividends in additional fuel capacity, a better fineness ratio for enhanced transonic acceleration, and the possibility of locating four Sky Flash air-to-air missiles in semi-recessed positions in the lower fuselage. This example also sports a pair of Sidewinder short-range missiles on the inner sides of the pylons for the drop tanks.

RIGHT The General Dynamics F-16XL is clearly a close relative of the standard F-16 Fighting Falcon fighter, but the use of a 'cranked-arrow' delta wing and a 56-in/1.42-m longer fuselage boosts internal fuel capacity by 80%, allowing the type to carry twice the payload over the same combat radius as the F-16, or the same payload over double the combat radius.

The F-16XL's basic flight performance remains roughly comparable with that of the standard F-16; where it really comes into its own is in weapon carriage, with a maximum of 29 hardpoints available on 17 stores stations. The stores stations are semi-conformal types to minimize drag, which is 58% less than that of the F-16, enabling the F-16XL to carry twice the payload 45% further than the basic F-16. The conformal carriage of weapons is already becoming important. External carriage of disposable weapons became common in World War II. It allowed obsolescent fighters to be fitted with bombs and rockets for ground-attack work, and current types to carry drop tanks as a means of extending range, and weapons have been carried externally ever since. It was realized all along that external stores produced a high level of drag, to the detriment of performance, but operational flexibility was deemed more important than the reduction of drag by the elimination of exterior pylons and their fittings. Some dedicated single-role aircraft, such as the B-52 heavy bomber and the Convair F-102 and F-106 interceptors, had internal weapons bays, but in general the accommodation of weapons inside the fuselage was deemed unnecessary.

MIKOYAN MiG-21 FISHBED-K

TYPE: *single-seat dual-role fighter and light attack aircraft*
WEIGHTS: *empty 12,302 lb/5,580 kg; maximum take-off 20,723 lb/9,400 kg*
DIMENSIONS: *span 23 ft 5½ in/7.15 m; length 51 ft 8½ in/15.76 m including nose probe; height 13 ft 5½ in/4.10 m; wing area 247.6 sq ft/23.0 m²*
POWERPLANT: *one 14,550-lb/6,600-kg afterburning thrust Tumanskii R-13-300 turbojet*
PERFORMANCE: *speed 1,385 mph/ 2,230 km/h; ceiling 50,030 ft/15,250 m; range 460 miles/740 km*
ARMAMENT: *one 23-mm GSh-23L twin barrel cannon and up to 3,307 lb/1,500 kg of disposable stores (including four AA-2 Atoll or AA-8 Aphid short-range air-to-air missiles, cannon pods, rocket pods and free-fall bombs) on four underwing hardpoints*

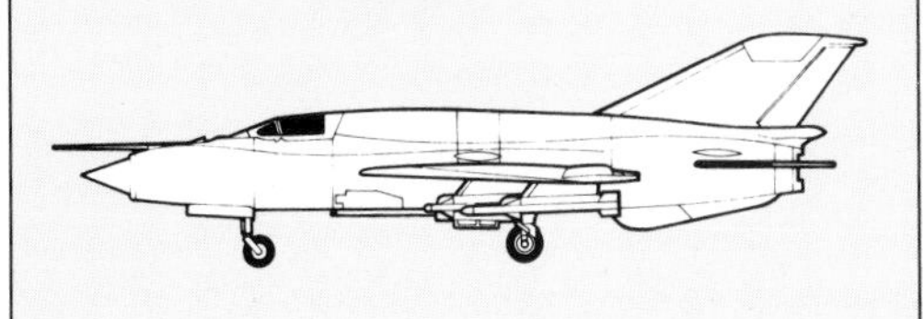

LEFT Considerable attention is devoted to the maximum use of internal volume, the Common Strategic Rotary Launcher being a good example; it allows the carriage of eight weapons such as these AGM-86B air-launched cruise missiles, which remain trimly folded until they have been launched.

HUGHES AIM-54 PHOENIX

Undoubtedly the most capable air-to-air missile in the world, the Hughes AIM-54 Phoenix is carried only by Grumman F-14 Tomcat fighters operated by the US Navy, the similar missiles and aircraft supplied to the Imperial Iranian Air Force being no longer serviceable. The missile was designed in parallel with the Hughes AWG-9 radar fire-control system for installation in the General Dynamics F-111B naval fighter, and when the overweight F-111B was cancelled became the basis for Grumman's successor design. The AWG-9 radar has a look-down search range of 150 miles/241 km or more and can track multiple targets, and the Phoenix was optimized to match the radar's capability.

Weighing 985 lb/447 kg at launch, the Phoenix has a speed in excess of Mach 5 and a range in excess of 125 miles/201 km, performance provided by a long-burning Aerojet Mk 60 or Rocketdyne Mk 47 solid-propellent rocket. After launch the missile climbs to about 100,000 ft/ 30,480 m and attains a burn-out speed of Mach 3.8 under the control of an autopilot which steers the missile towards the anticipated target position. After rocket burn-out, the missile slowly descends and accelerates to Mach 5+: as it does so, it acquires the target using its radar in the semi-active mode to receive electromagnetic energy bounced off the target by the launch aircraft's radar.

As the missile homes on the target its radar switches to active mode, transmitting its own signal and using its antenna to determine the target's exact position and range. For its size the Phoenix is remarkably agile even at long range, since the control fins aft of its delta wings are hydraulically powered. The warhead is a devastating 132 lb/60 kg unit carried just ahead of the wings; it is initiated either on impact or at the optimum distance from the target by a Downey Mk 334 radar or Bendix infra-red proximity fuse.

The initial AIM-54A variant introduced in 1973 was superseded in 1977 by the AIM-54B, which has a simplified structure, wings of sheet metal rather than honeycomb and non-liquid hydraulic and cooling systems. Development is currently under way of the AIM-54C model: matched to the latest Tomcat variant, this features an upgraded Nortronics interial reference unit, improved electronic counter-countermeasures and solid-state digital electronics in place of the earlier variants' klystron-tube analog electronics.

NASA/USAF
NASA
HiMAT

USAF/NASA
HiMAT
NASA

In the late 1960s the trend began to swing the other way with the semi-recessed or palletized carriage of missiles on aircraft such as the F-14 Tomcat, and similar arrangements are evident on the new generation of combat aircraft, especially those with an air-combat and/or air-superiority role for which minimization of drag is very important. This tendency is certain to continue, and may lead to the reintroduction of internal weapons bays for small air-to-air missiles carried in large numbers.

CORE AIRCRAFT

The process of treating the fuselage and wing roots as a structural core to which can be attached an evolutionary series of flying surfaces has been evaluated by the Rockwell HiMAT (Highly Manoeuvrable Aircraft Technology) remotely piloted vehicle of the early 1980s. The programme failed to explore the full range of possibilities, but the scope of the original concept emphasizes the way evolutionary development of core aircraft could proceed. Using the core as the structural, electronic and propulsive foundation, the HiMAT was flown in a canard layout and was schemed with a planar arrow wing, variable-incidence wing and forward-swept wing, and was also envisaged with a two-dimensional vectoring-thrust nozzle on its engine.

Evolutionary design of this type makes excellent economic as well as aerodynamic sense. Before the mid-1950s the peacetime service life of a combat aircraft could be measured in months, and was far less in war. Since then the cost of developing and manufacturing each type has escalated so enormously that the designer has to consider a service life of 15 or more years before a completely new design can be perfected for service and actually afforded by the operator arm. It is impossible for even the most skilled design team to predict the range and type of technological improvements that may become available in that time, yet it is essential that such improvements should be incorporated whenever possible to prevent a potential enemy from securing any decisive technological advantage.

OPPOSITE, ABOVE The Rockwell HiMAT is one of the most fascinating of modern research tools into aerodynamic and aircraft structures. It is a remotely-controlled miniature aircraft capable of supersonic performance. The modular design allows the machine to be configured in several ways on a standard core section.

OPPOSITE, BELOW The HiMAT in flight.

BELOW The HiMAT has allowed the investigation of flight at very high angles of attack, with a close-coupled canard configuration.

GRUMMAN F-14A TOMCAT

TYPE: *two-seat carrier-borne fleet air-defence fighter with variable-geometry wings*
WEIGHTS: *empty 40,104 lb/18,191 kg; maximum take-off 74,348 lb/33,724 kg*
DIMENSIONS: *span 64 ft 1½ in/19.54 m spread and 38ft 2½ in/11.65 m swept; length 62 ft 8 in/19.10 m; height 16 ft 0 in/4.88 m; wing area 565 sq ft/52.49 m² spread*
POWERPLANT: *two 20,900-lb/9,480-kg afterburning thrust Pratt & Whitney TF30-PW-412A turbofans*
PERFORMANCE: *speed 1,545 mph/2,486 km/h; ceiling 50,000+ ft/15,240+ m; range 2,000 miles/3,220 km*
ARMAMENT: *one 20-mm multi-barrel cannon and 14,500 lb/6,577 kg of disposable stores (generally four Phoenix long-range, two Sparrow medium-range and four Sidewinder short-range air-to-air missiles) on four underfuselage pallets and two underwing hardpoints*

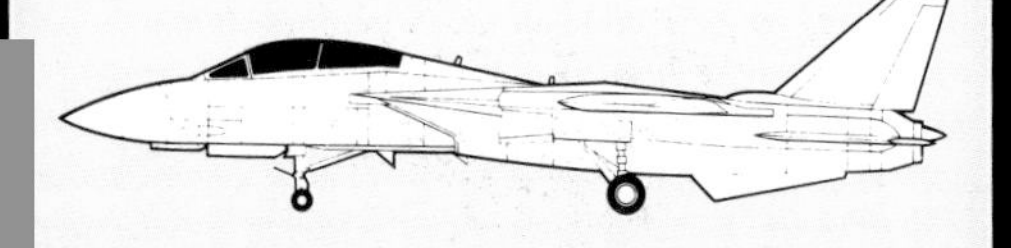

ABOVE The Grumman F-14A Tomcat has been dogged throughout its life by powerplant deficiencies, but is still a type that attracts superlatives in the description of its overall design. This example carries two AIM-7 Sparrow and two AIM-9 Sidewinder missiles under the outer ends of its fixed centre section, leaving the underfuselage positions possibly occupied by four more Sparrows or more probably a quartet of AIM-54 Phoenix long-range missiles.

LEFT The F-14s Tomcat, foreshadowed in this Super Tomcat prototype, introduced digital electronics and other improvements, plus a considerably improved powerplant for greater flexibility of operation.

RIGHT The Tomcat in its operational regime, providing US Navy carriers with long-range defence against air and missile attack.

ABOVE The large size of the AIM-54 Phoenix is readily apparent as the deck crew of USS *America* prepares a Tomcat for a sortie.

RIGHT The Tomcat has maximum flexibility of air superiority firepower, as revealed on this example which carries two short-range AIM-9 Sidewinders, two medium-range AIM-7 Sparrows and at least two long-range AIM-54 Phoenixes. There is also a 20-mm Vulcan cannon in the port side of the forward fuselage.

ABOVE RIGHT Tomcats on the flight deck of USS *Constellation* with their wings fully swept to reduce deck area requirement.

Consequently, the overall thrust of aircraft design in recent years has switched away from the concept of the aircraft as a throw-away item to be replaced as soon as a new model becomes available, toward the notion of the core airframe and powerplant as the basis for continuing development. Already, combat aircraft are designed with a mid-life update in mind, using modular features and databuses to allow the installation of more advanced components. The F-14D version of the Tomcat, for example, will be an altogether more formidable warplane than the current F-14A. New-generation digital electronics will replace the original analog electronics, upgraded weapons will be matched to the new electronics, and new engines will provide 30% more power plus far greater flexibility and reliability.

NEW MATERIALS

Aerodynamics are ineluctably tied to structure: it is good news when the design team's aerodynamicist comes up with the right answer to a requirement, but unless the structural designer can turn this answer into hardware it is of no practical value. Wood was the first primary structural medium for aircraft, but it rapidly faded from use in advanced aircraft as aluminium alloys came to the fore. Such alloys had been used experimentally even before World War I, together with steels of various types, but from the early 1920s the advantages of aluminium alloy as the primary structural medium dictated the development of powered aircraft.

At first aluminium alloys were used merely as a substitute for wood, aluminium alloy tubes simply replacing wooden longerons and spacers in a primary structure that was internally braced with wires and then covered with fabric or plywood. Then the implications of semi-monocoque construction were fully digested and the designers realized that an aluminium alloy skinning, suitably reinforced on its inner side by stringers and frames, could be used for combined aerodynamic and structural purposes. The resulting elimination of much of the internal structure and bracing that had increased structural weight and reduced usable internal volume, as well as making it difficult to produce effective cantilever structures, paved the way for the modern aircraft with its markedly enhanced performance and features such as retractable landing gear, high-lift devices and enclosed cockpits that have become standard.

Aluminium alloys continued to dominate the structural scene until the early 1980s. There have been exceptions, but mainstream aircraft production remained faithful to aluminium alloys of one kind or another up to the point at

GEODETIC CONSTRUCTION

Though no longer used, the geodetic system of construction enjoyed considerable popularity in the 1930s and early 1940s. Developed largely by Sir Barnes Wallis for a series of airships built by Vickers it takes its name from the geodetic line, the shortest line between two points on a surface.

Geodetic construction involves the creation of curved space frames whose individual members follow geodetic lines along the surface, each undergoing either compression or tension. The resultant structure looks somewhat like open-weave basketware, but has the advantages of great strength at modest weight. As well as creating a truly fail-safe structure it obviates the need for a stressed-skin covering and, in the context of the 1930s, was relatively straightforward to build.

The structure of individual members was pioneered in the disastrous M.1/30 biplane torpedo bomber and a primitive geodetic structure was used for the G.4/31, another biplane bomber. The latter had four light-alloy longerons over which were wound clockwise and anti-clockwise spiral channels to create a light yet very strong lattice structure.

The first aeroplane with a proper geodetic structure was the Wellesley monoplane bomber, the type used to set a world distance record of 7,158.444 miles/11,520.42 km between Ismailya in Egypt and Darwin in Northern Australia between November 5 and November 7, 1938. But the ultimate expression of the geodetic structure was the Wellington bomber of World War II.

In the Wellington Wallis refined his concepts into an immensely strong structure that ultimately proved simple to make once the semi-skilled work force had grown accustomed to the required techniques. Subsequent Vickers bombers built with a geodetic structure were the modestly successful Warwick, designed as a heavier counterpart to the Wellington, and the unsuccessful Windsor four-engined heavy bomber.

which aerodynamic heating began to enter the picture. Other materials were used in lightly-loaded areas, and glass-reinforced plastics (GRP) became particularly popular. GRP is used for complex shapes where light weight was desired, such as wingtips and fairings in areas of minimal aerodynamic tension, and for coverings over electronic installations such as radomes.

FIBRE-REINFORCED COMPOSITES

From GRP has been developed a new structural medium that may well prove as significant as aluminium alloys in the evolution of aviation technology. This is fibre-reinforced composite (FRC), which comes in a variety of forms that can be tailored to exact and exacting technical specifications. FRC is essentially GRP further reinforced with a layer or layers of lightweight carbon-graphite or boron fibres of exceptional stiffness and strength. Unlike a metal sheet of uniform thickness, which flexes equally in two planes, FRC can be made undirectionally stiff: the fibres can be laid in such a fashion that the resultant sheet of FRC material will flex in one plane without difficulty, but remain rigid in the other plane.

FORWARD-SWEPT WINGS

FRC materials have opened the way for the forward-swept wing, whose advantages have long been recognized by aerodynamicists. The first such wing was flown as far back as 1944 in the Junkers Ju 287 research aircraft intended as a prototype bomber. As the Germans discovered, however, the structural materials of the day could not prevent the phenomenon of aeroelastic divergence: any sudden control movement can cause the wings to flex, causing one wing's angle of incidence to increase, thereby generating additional lift (structural load) and yet more divergence (steadily increasing structural load) in a rapid cycle that results in the wing being torn off in a few fractions of a second by uncontrollable aerodynamic forces. Swept-back wings do not suffer from this structural problem, since the load rapidly reduces and damps rather than increases the aeroelastic divergence.

Nevertheless, the advantages of swept-forward wings are theoretically enormous. They delay any rise in transonic drag, reduce stalling speed, enhance low-speed handling

ABOVE AND LEFT The de Havilland Mosquito was an early exponent of composite construction, though in this instance the composite was a sandwich of balsa between outer layers of plywood. The result was a smooth, tough and easily repairable structure that was light and made minimum demands on the strategic alloys that were in short supply during World War II.

characteristics and provide virtually stall-proof flight. In combination these features should make it possible for a smaller aeroplane to have greater performance and agility on a given level of power, or the same performance plus greater agility on less power. The advantages in economic and tactical terms are enormous, and the opportunity offered by FRC has been widely explored, principally by the Grumman X-29.

The X-29, which first flew in December 1984, is purely a research tool, and uses the forward fuselage and cockpit of a Northrop F-5 fighter to reduce development costs. A design of relaxed static stability, it is powered by a General Electric F404 afterburning turbojet for a maximum speed of Mach 1.6+. The type's *raison d'être* is the forward-swept wing structure, which is attached about two-thirds of the way aft along the fuselage.

OPPOSITE, TOP A piece of skinning is lowered onto the wing of an F-15 fighter under construction by McDonnell Douglas. The panel is of a new aluminium-lithium alloy, offering comparable strength but lighter weight than conventional aluminium alloys.

OPPOSITE, BELOW The Grumman X-29 is an invaluable research aeroplane, pioneering the flight evaluation of a forward-swept wing.

RIGHT, BELOW AND BOTTOM Only the development of carbon-based composite materials has made possible the development of the X-29's forward-swept wing, which offers aerodynamic advantages over the aft-swept wing – but only if the structure can be tailored to avoid twisting.

The core of the wing structure is an electron-beam-welded box of titanium and light alloy, providing an exceptionally sturdy but generally conventional basis for the outer aerodynamic surfaces. The latter are single-piece upper and lower skins made of CFRP (carbon-fibre-reinforced plastics) up to 156 layers thick at their inboard ends. The skins are exceptionally light yet rigid, and can sustain the most violent aerial movements without any possibility of aeroelastic divergence. The leading edges are fixed, with no provision for high-lift devices of any kind, but the trailing edges are fitted with full-span flaperons that can be used as camber-changing sections. Flight tests have confirmed that these provide the forward-swept wing with very nearly the same mission capabilities as the Boeing MAW, with the added advantages of forward sweep and reduced mechanical complexity.

Located aft of the wing are the conventional rudder, plus a pair of strake flaps fitted at the extreme rear of the extended wing root trailing edges, nearly in line with the rudder. Powerful canard foreplanes with one-fifth of the wings' area are located on the sides of the lateral inlets, and just forward of the inboard sections of the wing leading edge, which are conventionally swept back. The canards are driven though a triply-redundant fly-by-wire flight-control system, and are the aeroplane's primary con-

LEFT The McDonnell Douglas AH-64A Apache is an inelegant but magnificent helicopter with good performance and highly capable electronics.

BELOW Complete with 16 missiles and underfuselage cannon, the Apache has a pugnacious appearance emphasized by the nose-mounted assembly of essential sensors.

RIGHT The attack helicopter is becoming increasingly important for the more sophisticated carriers, and the Apache is a leading contender in this field with the weapons displayed here: 30-mm rounds for the underfuselage cannon, 2.75-in/70-mm rockets for the multi-tube launchers, and up to 16 laser-homing AGM-114 Hellfire anti-tank missiles

trol surfaces in the pitching plane. In common with other canard surfaces, they are used to trim out any tail-up pitching moment by generating lift, augmenting the lift of the wing rather than reducing it in the fashion of tail-mounted pitch-control surfaces. The relationship of the canards and wings is mutually beneficial: the canards gain from an effective doubling of their moment arm, and the wings benefit from the downwash from the canards.

Flight tests have confirmed wind-tunnel predictions about the X-29's flight characteristics: even at extremely high angles of attack the aircraft cannot be stalled and it retains full roll authority down to very low airspeeds. Early flight trials also indicated that fuel burn was lower than expected, an indication of exceptionally low drag. There is every reason to forecast that the swept-forward wing made possible by FRC will become the norm for air-combat fighters, which will profit enormously from its increased agility. Attack aircraft should also benefit from the reduction in drag offered by the configuration.

Despite the success of the X-29 programme, current attention seems fixated on variable-camber wings as the primary means of optimizing the payload/range equation. There is no doubt that the MAW concept has great advantages over the conventional wing, but that is only achieved at the cost of mechanical complexity. Such complexity may be acceptable in a civil airliner, but is surely fraught with dangers in a military type, which must expect far more rigorous handling and the possibility of critical battle damage.

THE WAY FORWARD

FRC construction is certainly one of the most important ways forward in aircraft design, providing not just greater strength at reduced weight, but the possibility of tailoring that strength to secure specific aerodynamic advantages. As the technology matures, FRC is being used to replace aluminium alloys in new-generation combat and civil aircraft. In the longer term there is every reason to believe that more sophisticated FRCs will cause the very nature of aerodynamic design to be revised.

At the same time, there are new alloys such as aluminium-lithium, now being produced in bulk in the US, which will find extensive applications in civil and military aircraft. Recently, doubts have been expressed about the ability of FRC to cope with large-calibre cannon projectiles, but the manufacturers claim that the problems have been exaggerated and will be overcome as FRC matures in development and use. The largest FRC structure in military service is the wing of the Harrier II, a super-critical structure built of graphite/epoxy composite as a single unit.

HELICOPTERS

The helicopter differs from almost all fixed-wing aircraft in its ability to take off and land vertically and hover in the air. And while helicopters will never be able to match fixed-wing aircraft in any outright parameter of performance, their ability to translate at will between vertical and horizontal flight means that they can operate from confined areas such as jungle clearings and the heaving decks of small warships.

Helicopters began to appear in operational service in the last two years of World War II, but the value of early examples was limited by their marginal performance. The breakthrough in helicopter capability came with the turboshaft powerplant. Turboshafts are both smaller and more powerful than piston engines; they can be located close to the rotor assembly, simplifying and lightening the associated transmission system and leaving a clear volume for payload. Other advantages are greater power in overall terms, increased reliability and significantly reduced vibration levels.

It was the French who pioneered turboshaft-powered helicopters, followed closely by the Americans and then the Soviets. In its new guise the helicopter became a useful flying machine and an invaluable tool on the modern battlefield, rapidly evolving from the utility machine epitomized by the Bell UH-1 Huey into specialized types such as the light scouting helicopter, exemplified by the Hughes OH-6 Cayuse and dedicated gunships like the Bell AH-1 HueyCobra.

Trainable nose turrets for grenade-launchers, multi-barrel cannon or machine guns and hardpoints on stub wings for gun pods, rocket launchers and air-launched anti-tank missiles have become standard features on battlefield helicopters, as their potential has been realized.

At the other end of the scale are medium- and heavy-lift helicopters such as the twin-rotor Boeing Vertol CH-47 Chinook, the single-rotor Sikorsky CH-53 Sea Stallion series and the Mil Mi-26 Halo. These large machines are used to move troops and equipment on the battlefield and in support of amphibious forces.

The naval helicopter, meanwhile, has become a multi-role machine whose most important task – the detection and attack of submarines – is performed by such types as the Sikorsky/Westland Sea King and the Sikorsky SH-60 Seahawk. Naval helicopters can also attack small surface vessels using missiles such as the Kongsberg Penguin, British Aerospace Sea Skua and Sistel Sea Killer.

The growing electronic capability of the larger helicopters and their greater weight-carrying ability allow them to carry heavyweight anti-ship missiles such as the British Aerospace Sea Eagle and Aérospatiale Exocet, designed for use against large warships.

Developments are continuing apace, and modern helicopters such as the Westland Lynx combine a useful payload with excellent flight performance and aerial agility. The result is an increasing shift in emphasis toward fighter helicopters. Air-to-air versions of shoulder-launched surface-to-air missiles provide them with the ability to tackle heavier machines such as the Mil Mi-24 Hind and McDonnell Douglas AH-64 Apache attack helicopters.

ABOVE The composite wing of the McDonnell Douglas AV-8B Harrier II is light, large, and strong enough to support six underwing hardpoints for this versatile machine's wide array of armament options.

LEFT AV-8B construction, with the forward fuselage about to be mated to the aft fuselage before the installation of the vectored-thrust turbofan and the one-piece wing.

BELOW A ghosted illustration reveals some of the Harrier II's primary features: the black boxes contain electronics, the pale grey/blue colour shows the powerplant and thrust-vectoring system, the red highlights the fuel system, the middle blue reveals the reaction control system for control in the hover, the pale green marks the auxiliary power unit and the yellow is used for the air-conditioner/heat exchanger.

RIGHT The AV-8A is the US Marine Corps' equivalent to the British Harrier GR Mk 3, with the original small wing.

LEFT AND BELOW RIGHT The AV-8B: note the large number of design differences between this and the earlier AV-8A type.

CG
158390

2
POWERPLANT

All modern combat aircraft are powered by turbine engines of the type pioneered in combat during 1944 by the Gloster Meteor and Messerschmitt Me 262. The turbojet, universal until the mid-1960s, is a comparatively simple engine comprising compressor, combustion chamber and turbine, the last extracting just enough power from the gas flow from the combustion chamber to drive the compressor. Most of the power remains in the gas flow, which is expanded into the atmosphere at high velocity through a constricting nozzle and is often augmented by the afterburner. The afterburner – or reheat, as it is known in Britain – injects additional fuel into the exhaust: this fuel mixes with the free oxygen remaining in the exhaust gas and burns to create additional gas exhausted through a nozzle of larger area. The afterburner is effective only in supersonic flight; it can boost basic thrust by as much as two-thirds, but only by consuming disproportionate amounts of fuel.

The turbojet is still used for aircraft optimized for a single role or flight regime, whether subsonic or supersonic, though it is significant that aircraft powered by afterburning turbojets are those, like the MiG-25 interceptor and SR-71 reconnaissance aircraft, which are designed for

OPPOSITE, ABOVE A good example of how the powerplant conditions overall design is provided by this English Electric Lightning F Mk 6, whose fuselage accommodates a vertical pair of Rolls-Royce Avon afterburning turbojets, fuel and, large circular inlet.

OPPOSITE, BELOW In the years immediately after World War II airline operations were dominated by the USA's huge air-cooled radials, which offered great power, efficiency and reliability. This Lockheed L-749 Constellation has four Wright Double Cyclones.

ABOVE US Navy fighters such as this Boeing F4B-4 were largely responsible for the early development of the radial engine, in this case the Pratt & Whitney R-1340.

LEFT The Boeing 247 was the first 'modern' airliner, and was powered by two Pratt & Whitney Wasp radials.

MESSERSCHMITT Me 262 AND JUNKERS JUMO 109-004

The most significant combination of airframe and powerplant during World War II was that of the Messerschmitt Me 262 and its Junkers Jumo 109-004 turbojet. Junkers had started work on turbojets as early as 1936 and opted immediately for the axial-flow type rather than the less technically demanding centrifugal-flow type. The first Jumo 109-009 was running by the autumn of 1938, but the project and its staff were switched to Heinkel after Junkers' engine division discovered that jet engine development work was being undertaken in secret under the auspices of the company's airframe division.

The engine division of Junkers had already begun to move into turbojet development, and in the summer of 1939 it received a German air ministry contract for the new Jumo 109-004. This was designed for a thrust of 1,543 lb/700 kg at a speed of 559 mph/900 km/h, and was schemed round an eight-stage axial compressor, six combustors, a single-stage turbine designed with the aid of AEG's turbine expertise and provision for afterburning.

The first unit ran in November 1940, but was beset by so many technical problems that it was early 1942 before anything approaching reliability was attained; a Jumo 109-004A was flown for the first time under a Messerschmitt Bf 110 in March 1942, and other A-series engines were flown in prototypes of the Me 262. In the production-configured B-series the quantity of strategic materials was halved, largely through the replacement of castings by sheet metal constructions. This reduced both weight and the number of man hours involved in building each engine.

The Jumo 109-004B-1 first ran in May 1943, and the use of blades of improved shape in the first two stages of the compressor helped to increase thrust usefully to 1,984 lb/900 kg. This variant was quite rapidly cleared for production, the first units being delivered in March 1944 to power the classic Me 262A. Further development resulted in the Jumo 109-004B-4, with hollow rather than solid turbine blades, which entered production in December 1944; it was about to be supplanted by the 2,315-lb/1,050-kg thrust Jumo 109-004D-4 as the war ended. The Junkers plants and their work forces were seized by the advancing Soviets, and the whole production facility was moved to the USSR for continued development.

straight-line Mach-3 performance at high altitudes. Smaller turbojets are still used to power drones, remotely-piloted vehicles and sea-skimming anti-ship missiles.

THE TURBOFAN

The most common powerplant for modern aircraft, both civil and military, is the turbofan. This has the same basic core as the turbojet, but uses a far higher proportion of the gas flow from the combustion stage to drive the turbine: the turbine has more stages and drives a fan that both compresses the air for the core stage and ducts a substantial mass of propulsive air around the core section to generate the bulk of the thrust. For a given fuel consumption such an engine generates far greater thrust than a turbojet, especially at lower airspeeds, and is much quieter. The reduction in noise results from the cylinder of cooler air driven by the fan, which surrounds, smooths out and finally cools the hot and highly turbulent gas flow from the combustion chamber. The cooler air also reduces the engine's infra-red signature which forms the principal target for heat-seeking missiles. Like the turbojet, the turbofan can be augmented by an afterburner operating in the hot or cold gas flows.

The turbine has come a long way since World War II in terms of power output and reliability, while shrinking in weight and size and improving its specific fuel consumption – the unit of fuel required to generate a specific unit of thrust for a given period of time. For example, the Allison J35-A-35 of 1951 was 27 in/0.686 m wide and 195.5 in/4.97 m long, weighed 2,850 lb/1,292.76 kg, and produced an afterburning thrust of 7,500 lb/3,402 kg at a specific fuel consumption of 2.0. By 1965 matters had improved dramatically, the General Electric J79-GE-17 being 39.06 in/0.99 m wide and 208.7 in/5.30 m long, weighing 3,847 lb/1,745 kg, and delivering an afterburning thrust of 17,820 lb/8,083 kg at a specific fuel consumption of 1.97. Yet this achievement, producing one of the most widely used of all turbojet engines, is eclipsed by that of the Turbo-Union RB199 used to power the Tornado: only 34.25 in/0.87 m wide and just 127 in/3.225 m long, the RB199 weighs a mere 1,980 lb/898.1 kg while delivering an afterburning thrust of 16,000 lb/7,257.6 kg) with a specific fuel consumption of 1.5.

ABOVE AND LEFT The almost incredible Lockheed SR-71A 'Blackbird' strategic reconnaissance platform is powered by a pair of Pratt & Whitney J58 (JT11D-20B) bleed turbojets burning a unique fuel. These engines each develop a dry thrust of 23,000lb/10,430kg rising to 32,500lb/14,740kg with afterburning, and at high speeds develop much of their power as suction at the inlet.

McDONNELL DOUGLAS AGM-84 HARPOON PROPULSION

Missiles are generally associated with rocket motors of the solid-propellant type, but in recent years there has been a significant switch away from such propulsion systems. The reason is simple: solid rockets must carry both fuel and the oxidizer required to make it burn, limiting both the quantity of fuel and the missile's range. This hardly matters in short-range weapons such as dogfighting air-to-air missiles, for which rapid acceleration and instant power are more important. But where range is a principal requirement, as it is for sea-skimming anti-ship missiles, it is a severe handicap.

Sea-skimmers fly in the oxygen-rich lower reaches of the atmosphere, where it seemed pointless for them to carry their own oxidizer. The logical result was a new generation of subsonic missiles powered by small turbines: truly long-range weapons such as the Boeing AGM-86 air-launched cruise missile, with diminutive turbofans, and shorter-ranged weapons such as the AGM-84 Harpoon with one-shot turbojets.

The Harpoon's Teledyne CAE J402-400 evolved ultimately from the French Turboméca Marbore II, which was Americanized as the J69 as the start of Teledyne's mini-turbojet development programme in 1951. In the AGM-84 the J402 is rated at a sea-level thrust of 660 lb/299 kg turning at 41,200 revolutions per minute. A trim unit with precision-cast axial- and centrifugal-flow stages, it is located in the tail. Exhausting centrally between the rear control fins, it is fed with air via a ventral inlet and with fuel from a tank located above the inlet on the missile's centre of gravity, thereby avoiding shifts in the centre of gravity as the fuel is consumed.

FAIRCHILD REPUBLIC A-10A THUNDERBOLT II PROPULSION

One of the most interesting propulsion arrangements on a current combat aircraft is that of the Fairchild Republic A-10A Thunderbolt II. Designed for very modest overall performance, but requiring the ability to loiter for long periods over the battlefield, the A-10 relies on a pair of turbofans to produce the required endurance without excessive fuel consumption.

The engine involved, the General Electric TF34, was developed for the Lockheed S-3 anti-submarine aircraft, which also needs to be able to loiter for long periods at low levels. This useful engine has a single-stage fan and a 14-stage axial-flow compressor, is rated at 9,275-lb/4,207-kg thrust and weighs 1,478 lb/670 kg.

Just as important in the evolution of the A-10A was the engines' location on short pylons extending above and slightly out from the fuselage between the wings and tailplane. Such an installation is satisfactory from the aerodynamic point of view, offering minimal control problems in the event of loss of a single engine. More importantly, it means the wings, tailplane and much of the fuselage are between the ground and the engines, shielding the engines from ground fire and hiding them from ground-based sensors.

The flat face of the fan is the feature of turbofan-powered aircraft most readily detected by radar, and the A-10A's wing is below and forward of the engines to protect them from such detection. At the other end of the engines the tailplane conceals the exhaust plumes, whose signature is already reduced by the cold air propelled by the fan, from the gaze of infra-red homing missiles.

FAIRCHILD REPUBLIC A-10A THUNDERBOLT II

TYPE: *single-seat battlefield close-support and anti-tank aircraft*
WEIGHTS: *empty 24,960 lb/11,322 kg; maximum take-off 50,000 lb/22,680 kg*
DIMENSIONS: *span 57 ft 6 in/17.53 m; length 53 ft 4 in/16.26 m; height 14 ft 8 in/4.47 m; wing area 506 sq ft/47.01 m^2*
POWERPLANT: *two 9,065-lb/4,111-kg thrust General Electric TF34-GE-100 non-afterburning turbofans*
PERFORMANCE: *speed 439 mph/706 km/h; ceiling not relevant; range 576 miles/927 km with a 1.7-hour battlefield loiter*
ARMAMENT: *one 30-mm Avenger multi-barrel cannon and up to 16,000 lb/7,258 kg of disposable stores (including two or four Sidewinder short-range air-to-air missiles, six Maverick air-to-surface missiles, glide bombs with laser or electro-optical guidance, cannon pods, rocket pods and conventional or cluster bombs) on three underfuselage and eight underwing hardpoints*

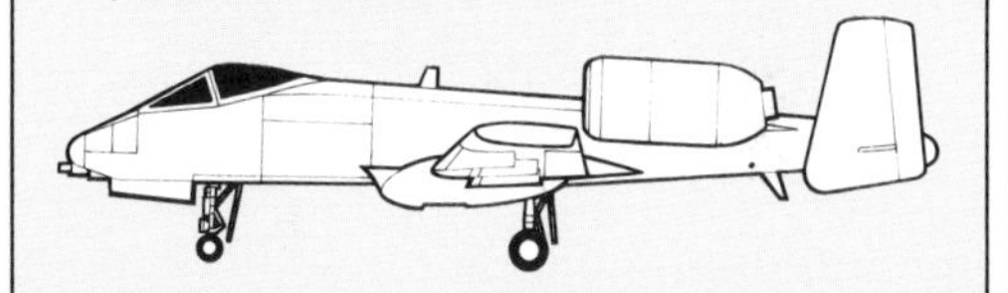

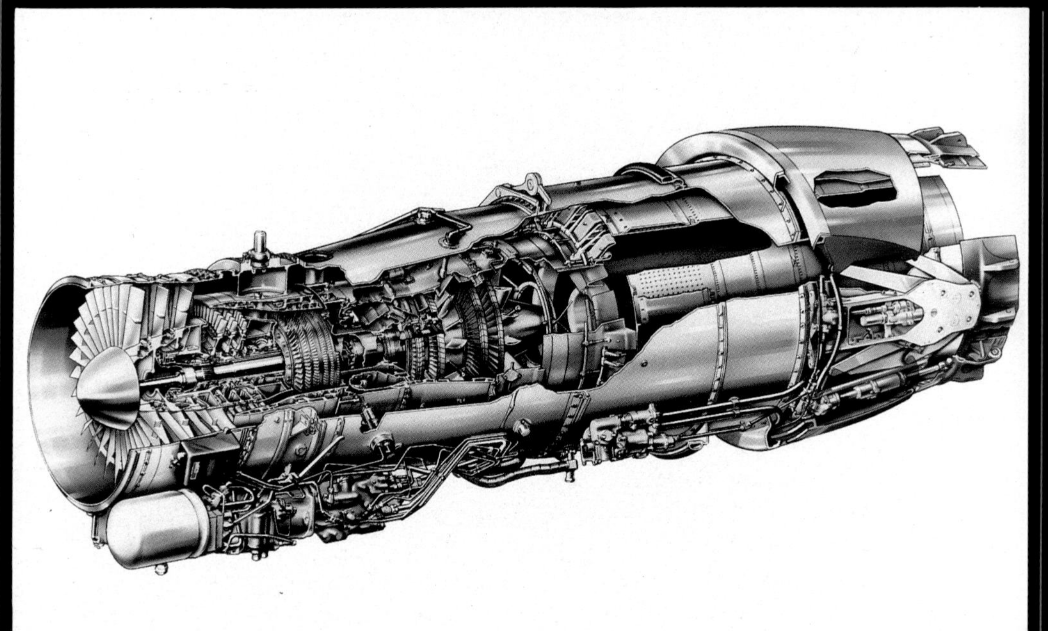

ABOVE The Tornado ADV's massive inlets provide striking evidence of the air quantities gulped by modern turbofans.

LEFT The Turbo-Union RB199 Mk 104 is a classic example of high power from an extremely compact design. Note the clamshell thrust-reverser round the nozzle.

ABOVE RIGHT Optimization of the RB199's thrust requires the use of fully-variable nozzles for the two engines.

ABOVE FAR RIGHT The Tornado ADV is an immensely impressive sight in full afterburner.

RIGHT The RB199 Mk 104D is used in the British Aerospace EAP research fighter.

6677

BYPASS RATIOS

Turbofans are generally assessed in terms of their bypass ratios – that is, the ratio between the cool air accelerated by the fan past the core section and the high-pressure air used in the core section. There is a great diversity between the turbofans used in combat aircraft and those that power transports. Combat aircraft must use an engine of small diameter, and this militates against the use of a large fan.

Typical turbofans used by current combat aircraft are the General Electric F404 (bypass ratio 0.34), Prat & Whitney F100 (0.7), Turboméca/Rolls-Royce Adour (0.8), Turbo-Union RB199 (1) and Rolls-Royce Pegasus (1.4). The higher the bypass ratio the better the specific fuel consumption is a general rule, and modern combat aircraft with high-bypass-ratio turbofans have significantly longer ranges than earlier aircraft on the same fuel load.

It is interesting to look at the evolution of the F-16's powerplant in the days when it was still the General Dynamics Model 401. Some thought was given to a powerplant of two General Electric J101 afterburning turbojets, which gave a mission weight – airframe, fuel and weapons – of 21,470 lb/9,739 kg. Then the concept was re-cast with a single Pratt & Whitney F100 afterburning turbofan, whereupon mission weight declined to 17,050 lb/7,734 kg, offering various useful possibilities. To give just one, the weight saving could be used for additional fuel, boosting range by more than 70%.

The range increment was attributable to the lighter powerplant installation (45%), lower fuel flow (40%), reduced airframe weight (11%) and reduced drag (4%).

Turbofans for fighters inevitably involve a compromise in design emphasis and effort, being designed to provide aircraft with both high performance and long range, as well as being optimized for extreme flexibility of operation so that the pilot can work the throttle as rapidly as required by the tactical situation. The high-bypass-ratio turbofan for long-range cruising performance is entirely different. Such engines are used in modern airliners and long-range military transports. Immediately noticeable for the diameter of their enormous fans, they have bypass ratios in excess of 4 allied to extremely good fuel consumption.

LEFT The nature of the Rolls-Royce Pegasus non-afterburning turbofan is readily apparent from this view showing the two cold-gas forward, and two hot-gas aft, thrust-vectoring nozzles.

BELOW LEFT Finnish technicians check a Turbomeca/Rolls-Royce Adour non-afterburning turbofan before its installation in a BAe Hawk trainer. The Adour has been developed in afterburning and non-afterburning forms, both notable for their good power-to-weight ratios.

RIGHT Clearly visible on this F-16 two-seater is the fully-variable nozzle for the Pratt & Whitney F100 afterburning turbofan, aspirated via an inlet under the forward fuselage to optimize airflow at high angles of attack.

BELOW A ghosted view of the two-shaft Adour shows (from front to rear) the two-stage fan, five-stage high-pressure compressor, annular combustion chamber, single-stage high-pressure turbine, single-stage low-pressure turbine and nozzle.

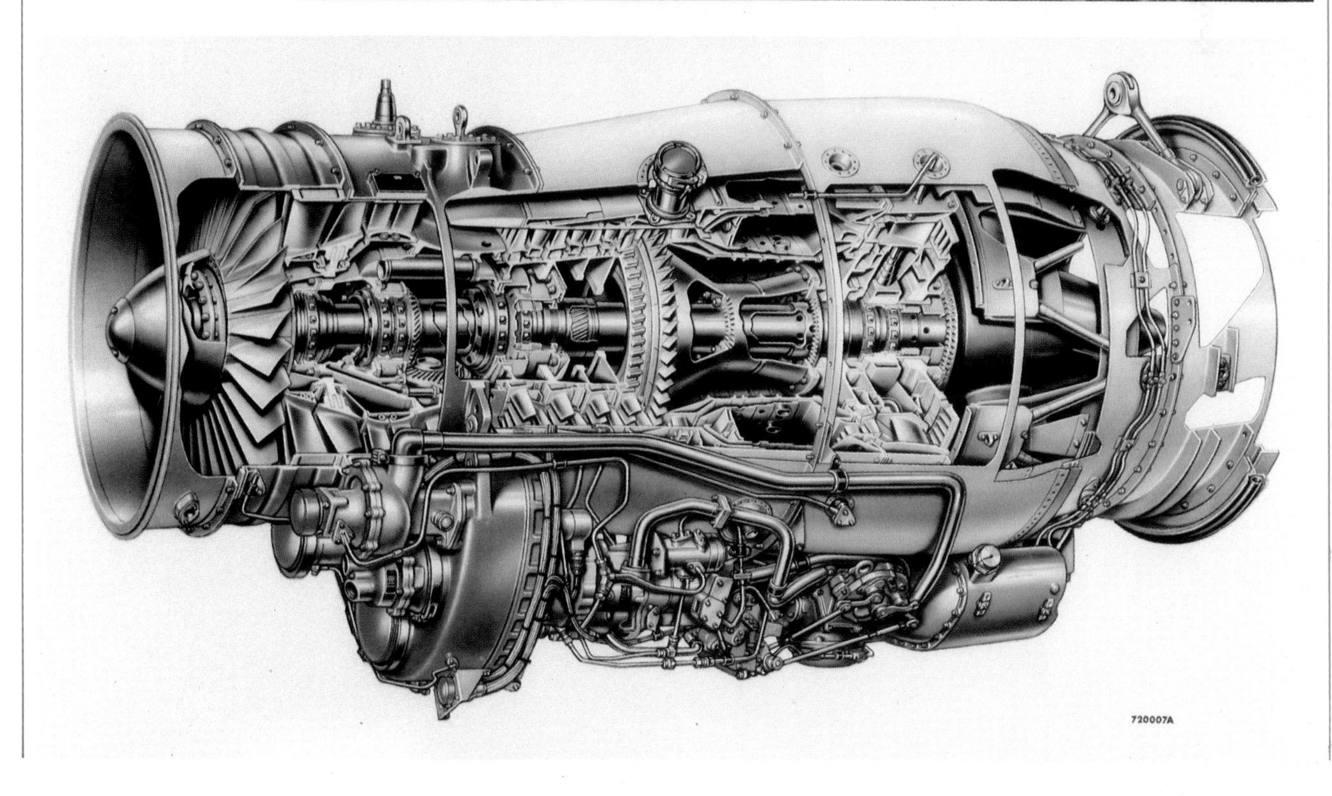

RIGHT Like the F-16, the McDonnell Douglas F/A-18A Hornet is optimized for modestly supersonic performance, though in this instance with a side-by-side pair of General Electric F404 afterburning turbofans.

BELOW LEFT The powerplant of the F/A-18A was designed for optimum operating capability and easy maintenance, the two F404s being located in 'straight-through' installations on the sides of the lower fuselage.

BELOW RIGHT Through retaining the standard F100 engine of its F-16B origins, the F-16XL has much enlarged fuel capacity.

OPPOSITE, ABOVE LEFT The F/A–18A has simple inlets, perfectly adequate for the type's modest performance.

OPPOSITE, BELOW The F/A-18A's powerplant is concentrated in the rear of the airframe.

OPPOSITE, ABOVE RIGHT The powerplant of the F/A-18A, like that of other modern aircraft, is voluminous: in red are the engines, and in yellow the fuel tanks.

CE15-2
151-02

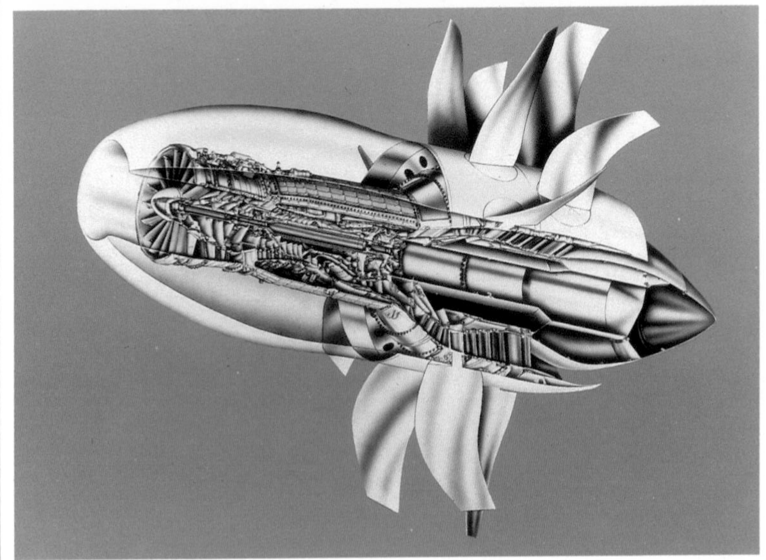

THE UNDUCTED FAN

A development of the late 1980s that seems set to complement the high-bypass-ratio turbofan is the unducted fan (UDF): this can be regarded as the modern equivalent of the turboprop, in which the main power of the turbine is used to drive a substantial propeller, but combines the turbofan's performance and fuel economy with the turboprop's low noise.

A characteristic of such engines, examples of which are under development in the USA by Allison, General Electric and Pratt & Whitney, is the use of an aft-mounted fan with multiple blades, made of aeroelastically stiff FRC and of swept configuration to mitigate the onset of compressibility problems as the tips approach Mach 1. Fears about the damage that might be caused by a blade separating from its fan mean that such engines are most likely to be tail-mounted. Most of the applications currently envisaged involve medium-capacity civil aircraft, but once the concept has been fully validated it is reasonable to expect that the UDF will be used for long-range miliary aircraft.

INLETS AND NOZZLES

Just as important as the specific engine is the arrangement of its inlets and nozzles. The nozzle is the region of the engine at which the hot gas flow is discharged into the atmosphere, its role being to convert as much as possible of the total energy of the gas into kinetic energy. The nozzle generally has a greater diameter than the engine itself, and is most commonly of the convergent/divergent type, in which the cross section converges into a throat where the subsonic gas flow is accelerated to supersonic speed, then diverges in order to allow further acceleration of the gas flow as it emerges into the atmosphere.

The convergent/divergent nozzle is fully variable to ensure that optimum conditions can be provided in all flight regimes, being constructed in a series of overlapping petals controlled by a powerful actuator system.

The design of the nozzle area is critical to the operation of the whole engine, and presents extreme aerodynamic difficulties as well as the expected metallurgical problems associated with an incandescent and vibration-charged plume of gas emerging from the engine.

The thrust of the engine can be used for

IN-FLIGHT REFUELLING

In-flight refuelling is such an integral part of air operations that it is easy to forget that the practice is a comparatively recent innovation for all but heavy bombers. Its most obvious use is to top up an attacking aircraft's tanks before it enters enemy airspace; but it can also enable aircraft to carry heavier loads by trading fuel for payload during takeoff and to sustain damaged aircraft.

There are now two basic techniques for in-flight refuelling, one using a flying boom, the other known as the probe-and-drogue type. The former involves a rigid boom controlled from the tanker: an operator uses the boom's aerodynamic controls to position its tip in the receptacle on the upper surface of the receiving aircraft, which formates precisely on a system of markings and lights on the tanker's underside. The flying boom system can be carried only by large aircraft with special tankage for rapid fuel transfer.

The hose-and-drogue system offers greater tactical flexibility. The tanker, which can be a combat aircraft fitted with a buddy refuelling pack drawing fuel from its own fuel system, trails a flexible hose fitted with a stabilizing drogue containing the female portion of the fuel-transfer coupling: the receiving aircraft flies into position below and aft of the tanker, and inserts its refuelling probe.

braking as well as acceleration. Most civil aircraft are fitted with a system of thrust reversal, normally in the form of cascades or clamshells to divert the thrust forward and so exert a powerful decelerative force.

Increasingly, however, the virtues of dispersed-site or damaged-runway operations are persuading the military powers that thrust reversal offers persuasive tactical advantages.

The most notable combat aircraft to incorporate thrust reversers are the Tornado and the Viggen, both exceptionally potent types: that of the Tornado is fitted upstream of the fully variable nozzle, and that of the Viggen allows the aeroplane to be flown straight onto the ground without a flare; as soon as the Viggen's nosewheel touches the ground and its leg is compressed the thrust-reversal system is automatically activated to bring the aircraft to a rapid halt.

OPPOSITE, TOP The General Electric UDF engine on test in the starboard position of a Boeing 727.

OPPOSITE, MIDDLE The UDF engine combines the best features of the turbofan and turboprop.

OPPOSITE, BOTTOM The Pratt & Whitney JT9D-7R4 is one of the world's most economical turbofans and is used on the Airbus A300-600 and A310, and on the Boeing 747SUD and 767.

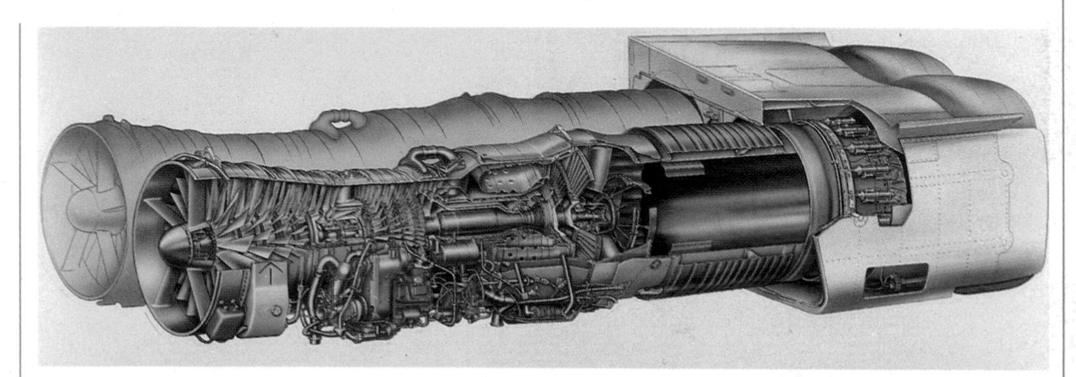

ABOVE RIGHT Still one of the most impressive powerplants in service, the Rolls-Royce/SNECMA Olympus 593 afterburning turbojet is batched in pairs under the wings of the Concorde supersonic airliner.

RIGHT The Rolls-Royce offering for large civil transports is the RB211, seen here in the form of the RB211-535C variant.

ABOVE The use of pylon-mounted pods for the engines of major airliners allows the purchasing airline to choose between several possible engines, the British Airways Boeing 757 being powered by a pair of Rolls-Royce RB211-535C turbofans.

LEFT Standard powerplant on the BAe 146 is a quartet of Avco Lycoming ALF 502R-5 turbofans, each rated at a thrust of 6,970lb/3,160kg. The use of four small turbofans increases reliability and reduces noise levels, allowing the BAe to operate by night into and out of airports otherwise closed to turbine-powered aircraft.

BELOW LEFT Typical of a military powerplant installation is that of the F-14A Tomcat fighter, with a pair of Pratt & Whitney TF30–P–412 afterburning turbofans in the rear fuselage.

ABOVE RIGHT The small size and fuel economy of the RB199 afterburning turbofan is a primary reason for the compact overall dimensions of the Panavia Tornado.

RIGHT Readily apparent in this illustration of a Tornado GR Mk 1 in flight are the buckets of the nozzle-mounted thrust-reversers in the stowed position.

630

ABOVE A unique STOVL powerplant arrangement is that of the Yakovlev Yak-38A, with a vectored-thrust Lyulka AL-21F turbojet in the aft fuselage and two directlift Koliesov ZM turbojets in the forward fuselage.

BELOW Yak-38 combat aircraft on the flightdeck of the Soviet carrier *Minsk*.

STOVL NOZZLES

The Harrier's unique four-poster nozzle arrangement taps high-pressure cool air from the compressor for ejection through the forward pair of vectoring nozzles while hot gas from the combustion stage is ejected through the aft pair of vectoring nozzles. The Pegasus turbofan has proved remarkably successful in this role, but is now nearing the limits of its design potential in its current form.

YAKOVLEV YAK-38 FORGER-A

TYPE: *single-seat carrier-borne fighter and multi-role combat aircraft with STOVL (short take-off and vertical landing) capability*
WEIGHTS: *empty 16,281 lb/7,385 kg; maximum take-off 25,794 lb/11,700 kg for vertical take-off or 28,660 lb/13,000 kg for short take-off*
DIMENSIONS: *span 24 ft 0¼ in/7.32 m; length 50 ft 10⅓ in/15.50 m; height 14 ft 4 in/4.37 m; wing area 199.14 sq ft/18.50 m²*
POWERPLANT: *one 17,989-lb/8,160-kg thrust Lyul'ka AL-21F non-afterburning turbojet and two 7,870-lb/3,750-kg thrust Koliesov ZM non-afterburning turbojets, the former fitted with two vectoring nozzles and the latter being designed to operate purely as lift jets*
PERFORMANCE: *speed 627 mph/1,110 km/h; ceiling 39,370 ft/12,000 m; range 460 miles/740 km*
ARMAMENT: *up to 7,937 lb/3,600 kg of disposable stores (including AA-2 Atoll and AA-8 Aphid short-range air-to-air missiles, rocket pods, cannon pods, air-to-surface missiles and free-fall bombs) on four underwing hardpoints*

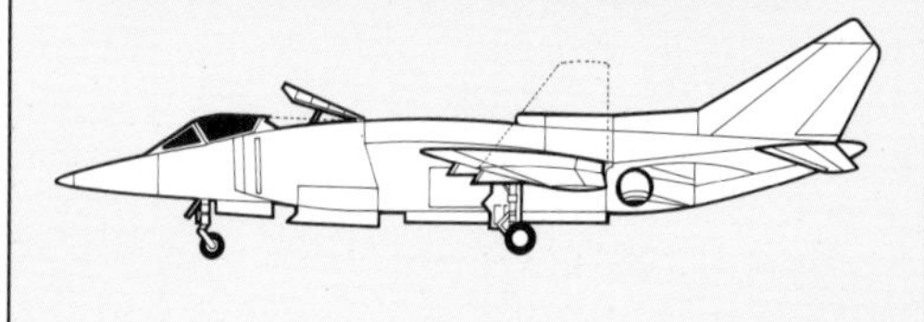

In the shorter term greater power can be derived from the introduction of plenum-chamber burning, a form of afterburning in the cool airflow before it emerges from the front nozzles. Greater long-term benefit may accrue from the use of a three-poster propulsion arrangement, which would allow plenum-chamber burning on the two lateral nozzles fed from the compressor, and afterburning on a single aft nozzle fed from the combustion stage. The tactical advantages of such a system would be considerable: greater thrust could be generated and drag would be reduced, while the afterburning would be maximized in efficiency by a straight-through jetpipe without the current pair of right-angle bends.

Although thrust-vectoring is currently used only on STOVL aircraft, it has great potential on more conventional aircraft. The exhaust of current conventional aircraft is directed through circular nozzles along or close to the aircraft's longitudinal axis to drive the aircraft forward. But in recent years designers have explored the possibility of using a two-dimensional nozzle arrangement to vector this thrust and so produce greater agility in combat aircraft. A new version of the F-15 is being prepared with rectangular nozzles that can be used to vector the thrust up or down: this will help to reduce the take-off and landing runs to a marked degree (in collaboration with canard foreplanes), but will also generate greater aerial agility by introducing a turning moment to the axial force currently provided by the exhaust.

The further possibility exists of vectoring the thrust laterally as well as vertically on widely separated engines such as those of the F-14, with differentially vectoring nozzles to aid control in roll. There is no way that thrust-vectoring could effectively replace aerodynamic control surfaces, but there is every reason to press ahead with the development of such a system to supplement the conventional flying controls of combat aircraft.

INLET EFFICIENCY

At the other end of the engine, the inlet system is the feature that decides the efficiency of the engine installation and therefore that of the aircraft's overall performance. Up to transonic speeds the inlet need not be complex, but thereafter its design becomes increasingly critical as speeds approach and pass Mach 2. The difficulty stems from the designer's natural desire to optimize the airflow to the engine in terms of mass, speed and configuration under all flight regimes. Whereas early jet fighters had simple inlets – plain circular or oval types on the North American F-86 Sabre and the MiG-15, or twin triangular units in the wing roots of the Hawker Hunter – supersonic aircraft featured inlets with fully variable areas and configurations, not to mention spill doors and auxiliary inlets, such as the lateral units on the F-4 Phantom II. Enormous ingenuity went into the design of such inlets, which reached

TOP Thrust-to-weight ratio is particularly important in the engines of VTOL aircraft, and considerable effort is being devoted to an upgraded Pegasus turbofan with plenum-chamber burning to boost thrust.

ABOVE The Pegasus is characterized by short length but considerable diameter.

their apogee on high-performance fighters such as the F-15, but it was recognized all along that they were both costly and complex, ultimately increasing aircraft weight and functional complexity.

Should the real utility of the inlet be doubted, it is sufficient merely to remember that on engines such as the SR-71's Pratt & Whitney JT11Ds and the MiG-25's Tumanskii R-31s some 70% of the thrust at maximum speed is provided by the inlets. Given the efficiency of modern inlets and nozzles, it is arguable that the ideal would be to replace the intervening turbine with an efficient ramjet to convert the inlet's offering into the gas needed by the nozzle. This is a nice idea, though it begs the question of how to power the aeroplane at speeds below the ramjet's effective operating speed (typically Mach 3).

The far-reaching analysis of combat aircraft undertaken by the US Air Force in the late 1960s and early 1970s to assess the implications of air warfare over Vietnam revealed how little time a Mach 2 fighter spent at Mach 2. The proportion may be only 0.1%, and while it may help the fighter to close on its prey before an engagement, Mach 2 capability is thereafter useless. Combat almost immediately degrades

LEFT A typical early-generation engine installation is evident on this Commonwealth Aircraft Sabre; the Rolls-Royce Avon non-afterburning turbojet is aspirated via a plain nose inlet and exhausts via a plain nozzle.

BELOW LEFT Altogether more complex is the arrangement of the F-15's two Pratt & Whitney F100 afterburning turbofans.

RIGHT An F-15 climbs in full afterburner.

BELOW The fuselage shell of the IAI Lavi multi-role fighter reveals the complexity of modern design and construction even before the addition of the engine and avionics.

BOTTOM The Lavi, after the installation of its Pratt & Whitney PW1120 turbofan.

performance to speeds of around Mach 0.8, at which the fixed inlet is more than adequate if properly designed.

Inevitably, therefore, the fully variable inlet has lost much of its attraction, and its elimination has simplified the task of the designer, manufacturer and maintenance man, as well as easing the burden on the taxpayer. Modern combat aircraft have plain inlets, often under the forward fuselage for optimum efficiency at high angles of attack. This means that speeds in excess of Mach 2 are no longer readily attainable, but the arrangement is more than satisfactory at speeds below Mach 2.

STEALTH TECHNOLOGY

Another feature of variable-geometry inlets that is a liability in the combat arena is their great radar cross section (RCS), the geometry of the inlets and their airflow-controllers involving an angularity that reflects radar energy beautifully. There is currently much emphasis on reducing the radar, electronic, thermal and acoustic signatures of combat aircraft. The whole field of stealth technology, as this science is known, is highly classified, and even the existence of a supposed stealth aircraft – the Lockheed F-19 reconnaissance fighter – is denied by the American authorities.

It is undeniable, however, that considerable strides have been made in the development of RAM (radar-absorbent material) technology

ABOVE Something of the electronic and systems complexity of modern combat aircraft is conveyed by this test rig, designed to evaluate the Lavi's environmental control and cabin pressurization systems.

LEFT, ABOVE RIGHT AND BELOW RIGHT The Boeing B-52G Stratofortress belongs to an earlier generation of combat aircraft, the modest power of contemporary turbojets requiring the use of eight such Pratt & Whitney J557-P-43 engines in four pairs under the wings, where their weight helped to reduce the bending on the wings. The powerplant system has remained essentially unaltered, but the B-52G's electronics have evolved through at least six phases to keep them as modern as possible, and evidence of this trend is seen in the forest of exterior antennae and fairings over the fuselage. The two blisters under the nose accommodate the low-light-level and forward-looking infra-red sensors of the variant's Electro-optical Viewing System for low-level penetration of enemy airspace.

low-RCS airframes. RAM is another highly classified subject, but the basic concept is comparatively straightforward: such materials are used in the structure or as coatings over the structure to absorb electromagnetic radiation. They enable the potential target aircraft to soak up the electromagnetic radiation broadcast by hostile radar systems, reflecting so little that the echo is difficult for the enemy to detect and then plot.

Inevitably, there are areas where the RAM is unsuitable. Here it is the designer's responsibility to reduce the reflective nature of the surface, or to build electromagnetic traps into it. A good contrast in such technology is provided by the US Air Force's heavy bombers, the Boeing B-52 Superfortress and Rockwell B-1B. The B-52 was designed in the late 1940s and early 1950s without any thought of reducing its RCS and has a massive slab-sided fuselage to

ABOVE LEFT The Rockwell B-1A was designed for supersonic penetration of enemy airspace at high altitude, and as such its powerplant installation of four General Electric F101-GE-100 afterburning turbofans was centred on variable geometry for optimized airflow conditions at all altitudes.

LEFT The production version of the B-1 is the B-1B low-level subsonic penetration bomber, and its F101-GE-102 afterburning turbofans are part of a simpler powerplant arrangement with fixed-geometry inlets but variable-geometry nozzles.

ABOVE Mock-up of the B-1's Offensive Avionics System operator station, with screens for the electro-optical viewing system (above) and radar (below), the latter controlled by the operator's joystick. In front of the operator is the alphanumeric screen and keyboard for interface with the system's computers, and to his left are the displays and controls for stores management and navigation.

which the wings, tailplane and rudder are butted at the maximum radar-reflecting angle of 90°. To make matters worse its engines are located in short nacelles that allow radars to see the compressor face of each engine without difficulty.

The B-1A was designed as the B-52's supersonic high-altitude successor, and achieved an RCS one-tenth that of the B-52 through the incorporation of blended contours and curved surfaces that reflected incoming radiation in a number of directions and so reduced that directed straight back at the emitter. When the B-1A metamorphosed into the B-1B, a low-level penetration bomber with lower speed, the original variable inlets were replaced by plain inlets with streamwise baffles to reduce the B-1B's RCS to a mere one-tenth of that of the B-1A – just one-hundredth of that of the B-52.

The design of the B-1B's inlets is a model for future development, the internal arrangement of the inlet and the baffle ensuring that radar cannot see the compressor face, the electro-magnetic radiation that enters the inlet being bounced about inside it and dissipated rather than reflected back to the emitter. The same basic approach has been taken with stealth aircraft such as the SR-71 strategic reconnaissance aircraft, whose very structure and design were conceived to reduce radar reflectivity: much of the SR-71's exterior surface is made of a RAM honeycomb, while its internal structure is arranged without right angles wherever possible, the use of small angles trapping the electro-magnetic radiation into a series of energy-sapping reflections inside the structure.

The new generation of combat aircraft feature an increasingly curvaceous outer appearance. The tendency is bound to increase as the advantages and techniques of stealth become more evident, and must soon involve the relocation of inlets in less obtrusive spots – perhaps as flush dorsal installations – as well as the adoption of internal weapon bays. At the same time, careful consideration of nozzles may help to reduce the acoustic and thermal signatures of modern combat aircraft, while the reduction of maximum speeds is also playing its part in reducing thermal signature by minimizing aerodymamic heating.

PULL TO EJECT

3
ELECTRONICS

The electronics associated with modern military aircraft are of extraordinary complexity and capability, and ultimately dictate what a particular aircraft type can achieve. The internal electronics are those associated with the aircraft's operation in concert with the pilot or crew, and external electronics are associated with its performance of its particular mission. This simplistic division serves to highlight the main tasks with which the electronics are entrusted.

Electronics are heavily implied in many of the features already discussed. Fly-by-wire control technology, for instance, is an extremely advanced form of digital processing using computer-written programs to deal with inputs from a variety of sources and allow the high-speed control of a wholly unstable aerodynamic platform in exact accord with the intentions of the pilot up to the point at which the airframe will become overstressed. These programs must be comprehensive yet flex-

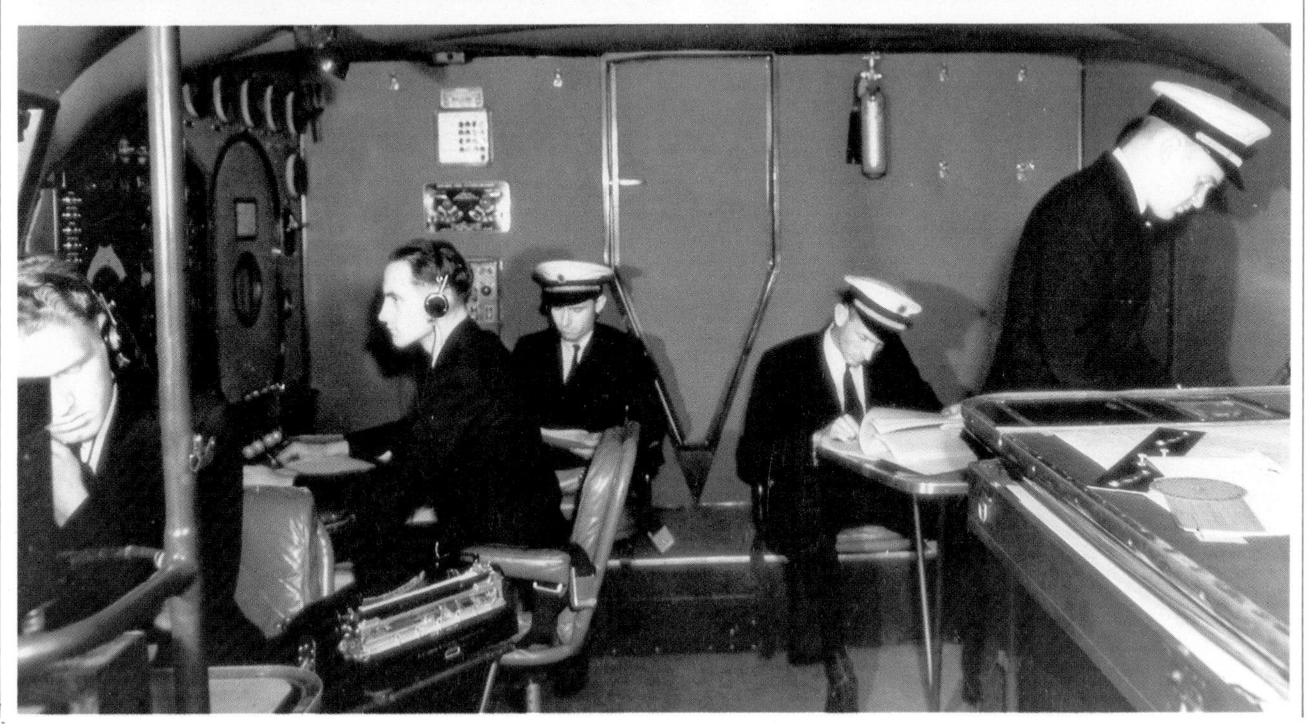

OPPOSITE, ABOVE The cockpit of the Douglas DC-4 is typical of piston-engined airliners in the late 1940s and early 1950s, with analog basic flight instruments, levers for the control of the throttles and propeller pitch, radio gear, trim wheels, fuel mixture and supercharger controls, and cowling and trailing-edge flap controls.

OPPOSITE, BELOW More spacious was the flightdeck of the Boeing 314 flying-boat airliner. This is its rear portion with a navigation table, radio position and spare seats.

OPPOSITE, MIDDLE The development of electronic flight controls and instruments in aircraft such as this SEPECAT Jaguar FBW paved the way for a cockpit revolution, with digital displays and controls.

LEFT The separate flight engineer's position on a Boeing 314.

BELOW The Jaguar's FBW's flight-control system was the UK's first quadruplex digital fly-by-wire system, allowing fully controlled flight under conditions of relaxed airframe stability.

ible, contain sophisticated algorithms (problem-solving routines), and feature a completely reliable system of making the right decision in each and every combination of factors. An analogous system is now becoming important for the optimized control of engines, ensuring that the powerplant functions efficiently in all circumstances.

These features are tied into the basic flight instruments – airspeed indicator, altimeter, horizontal situation indicator and attitude director – to create an autopilot that lifts the burden of routine flight from the pilot. Further sophistication of the system ties the autopilot into the inertial navigation system to produce extremely accurate navigation over long distances. And laser technology now allows mechanical gyroscopes to be replaced by lighter and altogether more reliable ring laser gyros.

COCKPIT DISPLAYS

The operation of the aeroplane's systems is also monitored in the cockpit. Until recently this required innumerable individual dials and gauges, and the pilot had to learn a complex series of routines to monitor the continuing progress of his flight in terms of aircraft and subsystem functions. This detracted from the time available for immediate tactical problems, and was an extremely tiring and boring process that tended to lull the pilot into a sense of false security.

This maze of dials and gauges was accompanied by a host of single-function switches which increased the complexity and unreliability of an already crowded cockpit.

Help for the pilot was found in strip rather than analog dials, so that each parameter could be displayed on a bank of vertical strip readouts: this allowed the pilot to cast his eyes rapidly over the instrument bank and see instantly if any of the strips was out of place.

The process was taken a step further with the introduction of cathode-ray tubes (CRTs), which allow a predetermined or selected number of parameters to be called up onto one or more screens as required. Combined with multi-function switches, CRTs have done much to simplify the cockpit without causing the pilot any loss of instrument information, especially as the screens and information technology have evolved to provide a more comprehensive yet more comprehensible presentation of data.

OPPOSITE, ABOVE The 'office' of the legendary Douglas DC-3.

OPPOSITE, BELOW Though designed on more modern ergonomic principles, the flightdeck of the BAe 146 is not conceptually far removed from that of the DC-3: it is more complex, but retains basic analog instrumentation.

ABOVE Well laid out with the minimum of clutter to distract the crew, the flightdeck of the McDonnell Douglas DC-10 marks an interim stage in flightdeck design, with analog instruments complemented by a small number of digital read-outs.

LEFT The flightdeck of the McDonnell Douglas DC–9 shows how comparatively small the crew's fields of vision are, except in the all-important horizontal and lateral directions.

RIGHT The cockpit of a fighter such as the F-15 Eagle is of necessity more cramped than the flightdeck of an airliner, but has many points of similarity in flight instrumentation. Peculiar to the fighter, of course, are features such as the stores-management system and radar display.

BELOW Careful design has now produced a flightdeck such as this for the DC-10, with primary flight instrumentation in front of each pilot, with engine, flap, landing gear and communications controls between them, and with additional switches on an overhead panel that can be reached easily by each member of the flight crew.

ABOVE The flightdeck of the DC–9, differing in scale and complexity rather than in substance from the flightdeck of the later DC–10.

LEFT One of the neatest 'conventional' flightdecks is that of the BAe Jetstream 31 commuter and executive transport.

The logical culmination of this process is the association of CRT displays with a computer. The computer is primed with all the standard readings, together with a number of diagnostic programs, and can alert the pilot with visual and aural warnings when any subsystem begins to stray outside the predetermined parameters. The computer can also detect incipient problems, which can then be cured or held in check before they affect the safety of the flight or the execution of the mission. The system also helps to reduce maintenance cost and time, with failing systems shut down before they suffer crucial damage, and a read-out available to the maintenance crew's diagnostic system.

In overall terms, then, the internal electronics add considerably to safety, improve the monitoring of subsystems and remove much of the drudgery of routine flight. There is a cost in terms of increased complexity – especially as the flight-control system must have inbuilt redundancy and the ability to monitor its own viability – and this has to be carefully considered by the design team. It is possible to add endless features, and the cost-effectiveness of the system has to be monitored closely. It is

OPPOSITE, ABOVE Since its arrival, the F-15 has been radically upgraded in many respects, not least in the cockpit. This now reflects the latest thinking, with single-function analog instrumentation replaced by a computer-controlled display system that automatically warns the pilot should any function stray beyond fixed parameters, and allows the pilot to call up any information he desires.

OPPOSITE, BELOW The cockpit of the F/A–18A Hornet is conceptually similar to that of the F-15 from the same manufacturer. The three main displays are the master monitor display (left) with the digital engine monitor display below it, the horizontal situation display (centre) and the multi-function display (right) with the radar warning display below and to the right of it.

ABOVE The pilot's cockpit of the F-14 Tomcat is of an older design generation, and dominated by the vertical situation display (above) and the horizontal situation displays (below). The radar and most other electronic systems are controlled by the rear-seater.

LEFT Another view of the F-14 pilot's cockpit, with the joystick and rudder pedals in the centre, and the throttles on the left.

0239

LEFT Lockheed Tristar flightdeck.

BELOW LEFT The 'bloodshot eyes' below the cockpit are the steerable sensors for the Westinghouse AVQ-22 low-light-level TV system (port) and Hughes AAQ-6 forward-looking infra-red system (starboard) of the ASQ-151 Electro-optical Viewing System of this B-52 bomber.

RIGHT The BAe 125–800 is the world's first business jet to offer an all-digital flightdeck as a customer option.

BELOW The two sensor turrets of the EVS fitted to the B-52H variant of the Stratofortress.

BOTTOM The less precise instruments of World War II aircraft demanded particular skills from the operator. This is a precision bomb-aiming sight aboard a Boeing B-29 Superfortress of the type that dropped atomic bombs on Hiroshima and Nagasaki.

also important that the pilot's functions should not be usurped completely: he must retain some functions or he may begin to feel redundant and fail to supervise the system effectively.

In military aircraft, the reason for this simplification of the pilot's task is easy to justify: the less time he has to spend with his eyes inside the cockpit, the more time he has for the primary task of assessing the external situation and coming to the right tactical decisions.

EXTERNAL EMISSIONS

Of greater tactical significance are the external electronics. These fall into two basic groups, active and passive. And just as radar cross section and other aspects of low observability have recently come to play a key part in the design of the airframe, the reduction of electronic emissions by active electronics is vital. They are as electromagnetically visible to passive electronic warfare systems as the beams of a powerful lighthouse to the human eye. Modern aircraft carry multiple emitters – single-beam main radar, terrain-following or terrain-avoidance radar, radar altimeter and tail-warning radar, plus four-beam Doppler navigation radar: each can betray the position of an intruder to the enemy searching for him, and in any successful engagement the detection of the enemy is the first step to his destruction.

Detection does not lead inevitably to destruction: the intruder may have several types of countermeasures, both active and passive. These can be carried as an integrated internal system, with antennas located in optimum positions, as a conformal system or as a podded installation. Conformal installations are becoming increasingly popular, since they offer the advantages of the podded type in ease of maintenance and upgrading while avoiding performance-degrading external installations and leaving hardpoints free for ordnance.

The most common passive countermeasure is chaff, which is composed of metallized plastic strips matched to the wavelengths of hostile radars and packed into dispensable cartridges. As soon as a threat radar is detected one or more chaff cartridges are fired into the airstream, where they explode and spread the chaff into a large cloud: this fills the enemy's radar screen with large numbers of false echoes, or alternatively decoys the radar astern of the intruder as it tracks the centroid of the decelerating chaff cloud. Conceptually akin to the chaff cartridge is the infra-red decoy, a specially designed flare constituting a heat source matched to the operating frequency of the enemy's heat-seeking missiles: this is ejected into the airstream to confuse or distract IR-homing missiles closing on the intruder.

ELECTRONIC COUNTERMEASURES

More sophisticated electronic countermeasures are either controlled by the parent aircraft's computer system or designed for autonomous operation. Such radar jammers usually operate in response to warnings from a passive radar-warning system by sending out a barrage of false returns to hide the real return, or by returning a slightly modified echo that conceals the target's true size, range and bearing. Other electronic countermeasures include wholesale jamming of the enemy's frequencies with broad-band noise transmitted at high power; and repeater jamming, in which the enemy's signals are retransmitted at a power inversely proportional to the received signal to cause inaccurate tracking in azimuth and elevation.

Electronic countermeasures is a growth area in modern military technology, and remains among the most highly classified subjects. Considerable strides have been made in this field over recent years, and the association of

Remotely-piloted vehicles are becoming increasingly popular for tactical reconnaissance. Their comparative cheapness and lack of human crew allows their dispatch into the most hazardous of situations, in which they stand a better chance of evading destruction because of their small size, and from which they can relay real-time information (both pictures and data) to a ground station. Typical of such craft are the Tadiran Master Mk III (on its launcher, **OPPOSITE, TOP**), the Tadiran Mastiff Mk III (**OPPOSITE, MIDDLE**), the Mastiff ground station (**OPPOSITE, BOTTOM**), the Canadair CL-89 battlefield reconnaissance drone (**ABOVE**) and the improved CL-289 on its launcher (**LEFT**).

GRUMMAN A-6E/TRAM INTRUDER

The Grumman A-6 Intruder was designed in the late 1950s to provide the US Navy with a carrier-borne all-weather strike and attack aeroplane. Using a state-of-the-art navigation system and multi-mode radar, the A-6 would deliver decisive loads of weapons over long ranges with unequalled accuracy. The prototype flew in November 1960, and the initial A-6A entered service in 1963.

The type soon proved itself in the adverse climatic and electronic conditions of the Vietnam War; it has been progressively upgraded to the current A-6E standard, with greater power, updated electronics and a larger diversity of weapon types.

The A-6E began to enter service in 1972, and from 1979 the stabilized Hughes TRAM (Target Recognition and Attack Multi-sensor) under-nose turret was introduced on production aircraft and retrofitted to older aircraft still in service.

The TRAM turret adds another dimension to the Intruder's all-weather capability, especially at the low levels now flown to avoid detection and engagement by hostile air-defence systems. A forward-looking infra-red sensor allows the Intruder's crew to see an extremely detailed thermal picture of the terrain ahead of or to the side of the aircraft. Once they have penetrated hostile airspace they can acquire targets visually for increased bombing accuracy. Meanwhile, a laser tracker-designator allows the aircraft to pinpoint a target for a laser-homing weapon or to deliver such a weapon against any target illuminated by another aircraft.

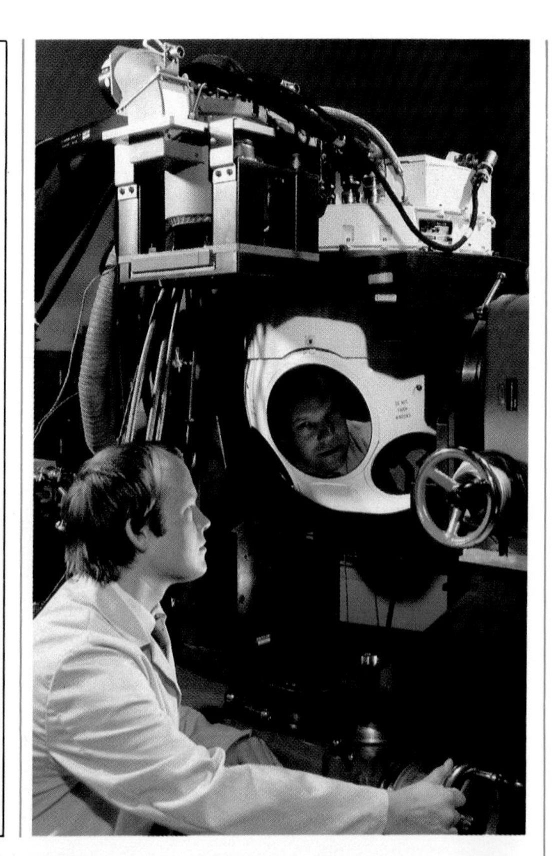

RIGHT Early electronic warfare pods, such as the QRC-335-3 on this F-4E Phantom in 1970, were designed for external carriage, allowing types to be changed between aircraft and also to be modified electronically to meet the evolving threat.

BELOW The Lockheed Aquila reconnaissance vehicle is launched by catapult from a special truck-mounted rail and recovered in a special net, into which the operator flies the vehicle.

electronic warfare techniques with powerful digital computers has done much to increase the sophistication of the techniques and the cleverness with which they are used for real-time disruption of the enemy's capabilities.

The role of the other external systems is largely self-explanatory. The main radar is used to search the volume of air ahead, while terrain-avoidance or terrain-following radars look forward and obliquely downward to detect obstacles in the flight path. Terrain-avoidance radar detects vertical obstacles round which the aircraft can be banked, while terrain-following radar allows the selected course to be flown at the selected height with just enough vertical variation to clear obstacles. The radar altimeter looks straight down to measure the exact height above the ground. And the Doppler navigation system provides dead-reckoning navigation by integrating along-track and across-track velocities derived from the Doppler shift in the returns from the ground of four diagonally-transmitted radar beams.

MULTI-ROLE RADARS

These are all key elements in the modern warplane's electronic suite, but the most important is clearly the main radar, which provides the pilot with most of his tactical information. Modern radar is extremely complex and capa-

LEFT The ALQ-131 is one of the US forces' most important tactical electronic warfare pods, and is a high-capability system designed to tackle the radars associated with surface-to-air missiles. The type has considerable growth potential, and is seen here having its software reprogrammed on the flight line.

BELOW LEFT Modern radars have an excellent mapping capability, these being images of the same airfield taken by the Hughes APG-65 radar of an F-15 (left) and by an optical camera.

BELOW The term 'chaff' is used for metallized strips dropped by aircraft to confuse the enemy's radar by filling its screens with a multitude of echoes. Modern chaff is normally loaded as rolls and cut to the desired length as it is released.

RIGHT Still one of the most capable in service, despite its considerable age, the radar of the F-14a Tomcat is part of the Hughes AWG-9 fire-control system matched to the mighty AIM-54 Phoenix air-to-air missile.

REMOVE BEFORE FLIGHT
F-602

GENERAL DYNAMICS/GRUMMAN EF-111A RAVEN

The most advanced electronic warfare (EW) aircraft currently operational, the EF-111A Raven is based on the General Dynamics F-111A variable-geometry strike and interdiction aircraft.

It carries an updated and more extensively automated version of the Grumman EA-6B Prowler's ALQ-99 tactical jamming system, computerized for operation by a single man rather than the Prowler's two-man tactical crew. And whereas the Prowler's jammers are carried externally in windmill-powered pods, limiting range and reducing range, the ALQ-99E's jammers are located semi-internally and powered by the aircraft's own power system.

The most notable external differences between the F-111A and the EF-111A are the long canoe fairing under the latter's fuselage and the large pod at the top of its fin. The ALQ-99E system uses the receivers in the fintop pod to detect hostile radar emissions, which are then passed to the computer system for analysis – of their frequency, identity and bearing – so that jamming frequencies and powers can be decided.

The jammers in the canoe fairing can defeat the world's most intense air-defence radar networks. So the Raven could be used to escort tactical aircraft formations, to open the way for interdiction aircraft or to disrupt hostile air-defence radars in a whole sector of the front.

ble, and whereas valve-technology radars were generally of a single type, requiring massive power and considerable cooling for their great bulk and weight, their modern equivalents are genuinely multi-role types based on the latest transistor and microchip technology of the solid-state variety.

This technology has allowed a considerable reduction in volume and weight, together with the associated power requirement, as well as considerably eased maintenance: built-in test equipment to detect and isolate faults is standard, allowing the maintenance crew merely to change the specified module without the need to remove the whole equipment. Apart from their logistical advantages, modern radars offer increased operating ranges, greater discrimination against surface clutter and glint, enhanced resistance to electronic countermeasures, the ability to track targets automatically while searching for new ones and in some instances the ability to illuminate the target with the electromagnetic radiation required by semi-active-radar-homing missiles.

The systems that have made all this possible are at the very forefront of modern technology, generally using digital signal processing to secure the sharpest possible image and to reduce false imagery. Such radars can be used in the air-to-air and air-to-surface modes, and in their look-down modes provide an increasing capability against small targets at very low levels. The radar is generally an integral part of the aircraft's fire-control system, providing the primary data needed for the solution of the incipient fire-control problem.

HEAD-UP DISPLAYS

Radar data can be displayed on a dedicated radar screen, on one of the cockpit CRTs, or on the pilot's head-up display (HUD). The HUD, which has become a fundamental part of cockpit equipment, enables alphanumeric

FAR LEFT The General Dynamics/Grumman EF-111A is an immensely capable and powerful electronic warfare aircraft, the antennae in its fintop fairing detecting the enemy's signals and feeding them to the computerized system that analyzes, identifies and localizes them before instructing the jammers in the ventral fairing to swamp them with their own signals.

ABOVE Pilots of military aircraft are now primed with important data via their head-up displays, onto which the symbology is projected in a form focused at infinity, so that the pilot can assimilate it without drawing his attention from the outside world. This is the HUD of an F-15 fighter.

LEFT The cockpit of the Harrier GR Mk 5 includes a head-up display (top), a moving map display (below, right) and multi-function display (below, left). This is the planning cockpit used to define the Harrier GR Mk 5's standards.

COMM 1
IFF
TCN
AWL
WPN
COMM 2
HUD SYM
STBY RET
ALT
DUCT
STAB
RWR
VOL
FUEL
BINGO
ECM

FAR LEFT The cockpit of the AV-8B is similar to that of the Harrier GR Mk 5, but has a second multi-function display instead of the moving map display.

LEFT Infra-red photography reveals features invisible to optical photography, as this image of a fuel tank farm by Texas Instruments RS-700 equipment shows. The white tanks are hot, indicating that they are empty, while the dark tanks are cold and thus full. Analysis can thus show the extent of the enemy's fuel reserves.

BELOW The McDonnell Douglas head-up display and control panel on the F-15 fighter.

data to be projected onto an inclined glass panel in front of the pilot's eyes. This means that he does not need to look down at his displays at crucial moments, the relevant data being presented in a stylized form focussed at infinity so that there is no need to refocus his eyes to assimilate the information. Early HUDs had only a very limited field of view – of the order of 13.5° left and right and 9° up and down – but current types typically provide angles of 20° and 15°, and wider fields are likely to be provided in the future.

The HUD can also accept data from laser rangefinders and the forward-looking infra-red (FLIR) systems. The function of the former is self-explanatory, providing the fire-control system with extremely accurate range and range-rate data of the type required for accurate gunnery and delivery of free-fall weapons. The type of laser is generally selected to operate in a frequency that will cut through the optical fog of a modern battlefield, and the installation can often be used to illuminate the target for a laser-homing weapon.

The FLIR provides another method of seeking through fog or night: great effort has been devoted to thermal imaging in recent years, and extremely detailed thermal images are provided by the latest equipment. The dual installation of radar and FLIR opens the possibility of continued target acquisition and target tracking even if the radar is jammed or in some

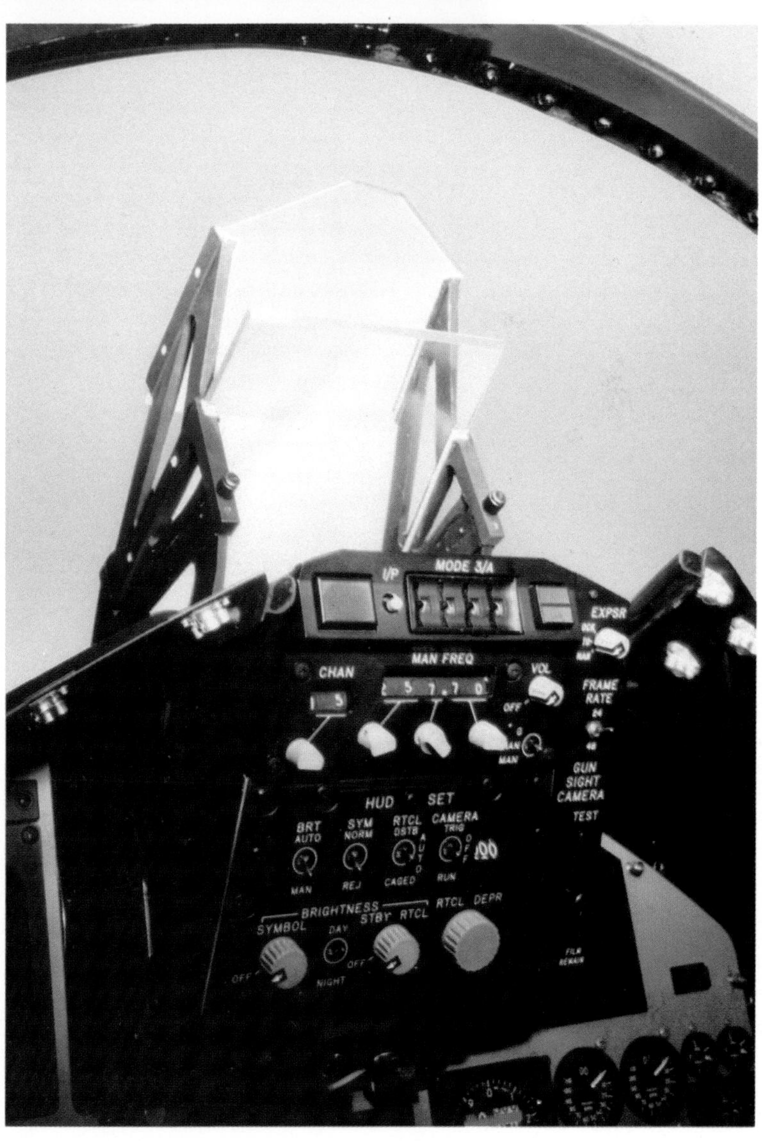

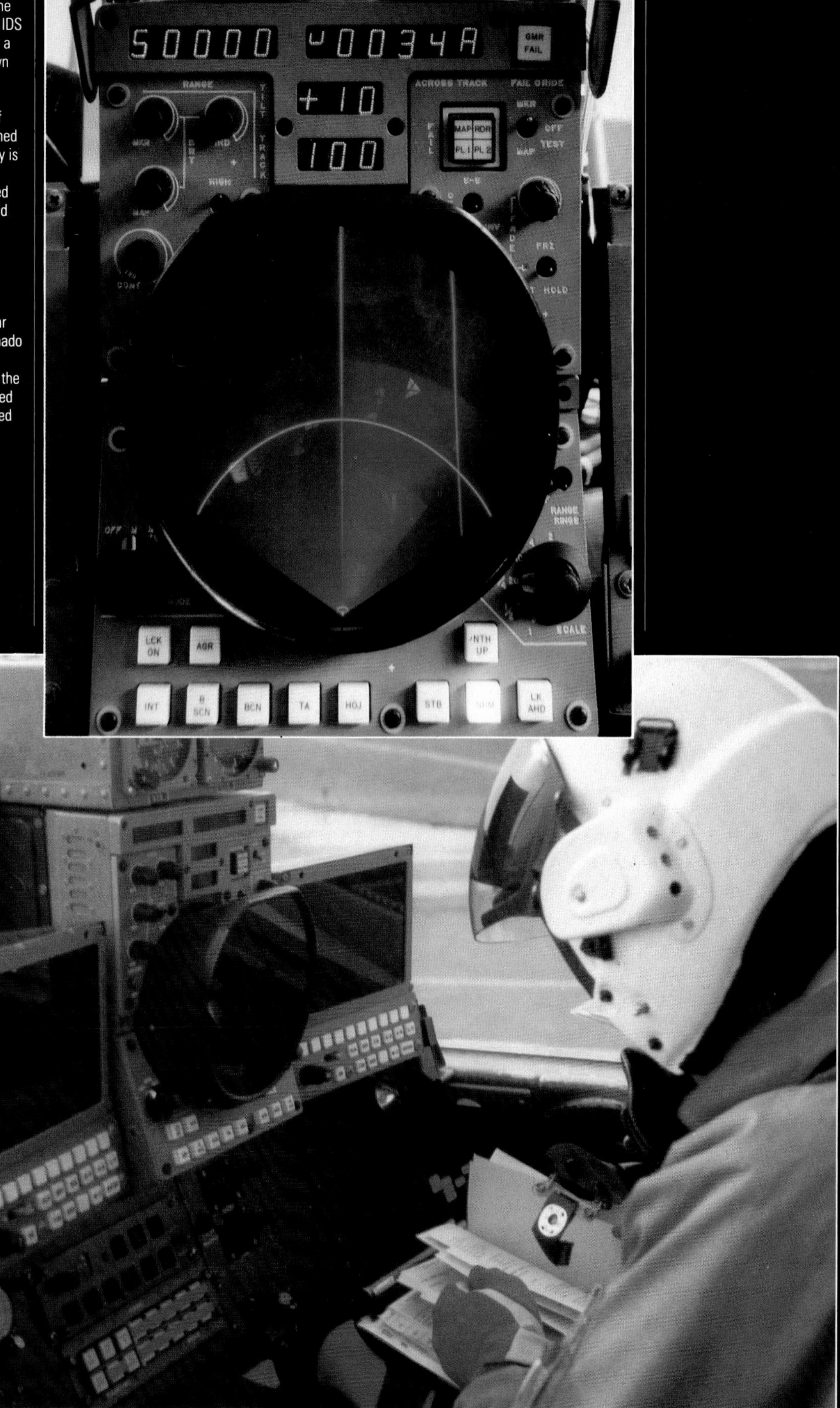

RIGHT The main display in the rear cockpit of the Tornado IDS can be used for radar or for a projected map, and is shown here in its radar mode.

BELOW In the rear cockpit of the Tornado IDS the combined radar/projected map display is flanked by a pair of multi-function displays and topped by an altimeter and airspeed indicator. The radar is controlled by the radar officer's hand controller.

RIGHT With the electronics controlled mostly by his rear seater, the pilot of the Tornado IDS can concentrate his attentions on flying and on the tactical situation; his centred circular display is a projected map display repeater.

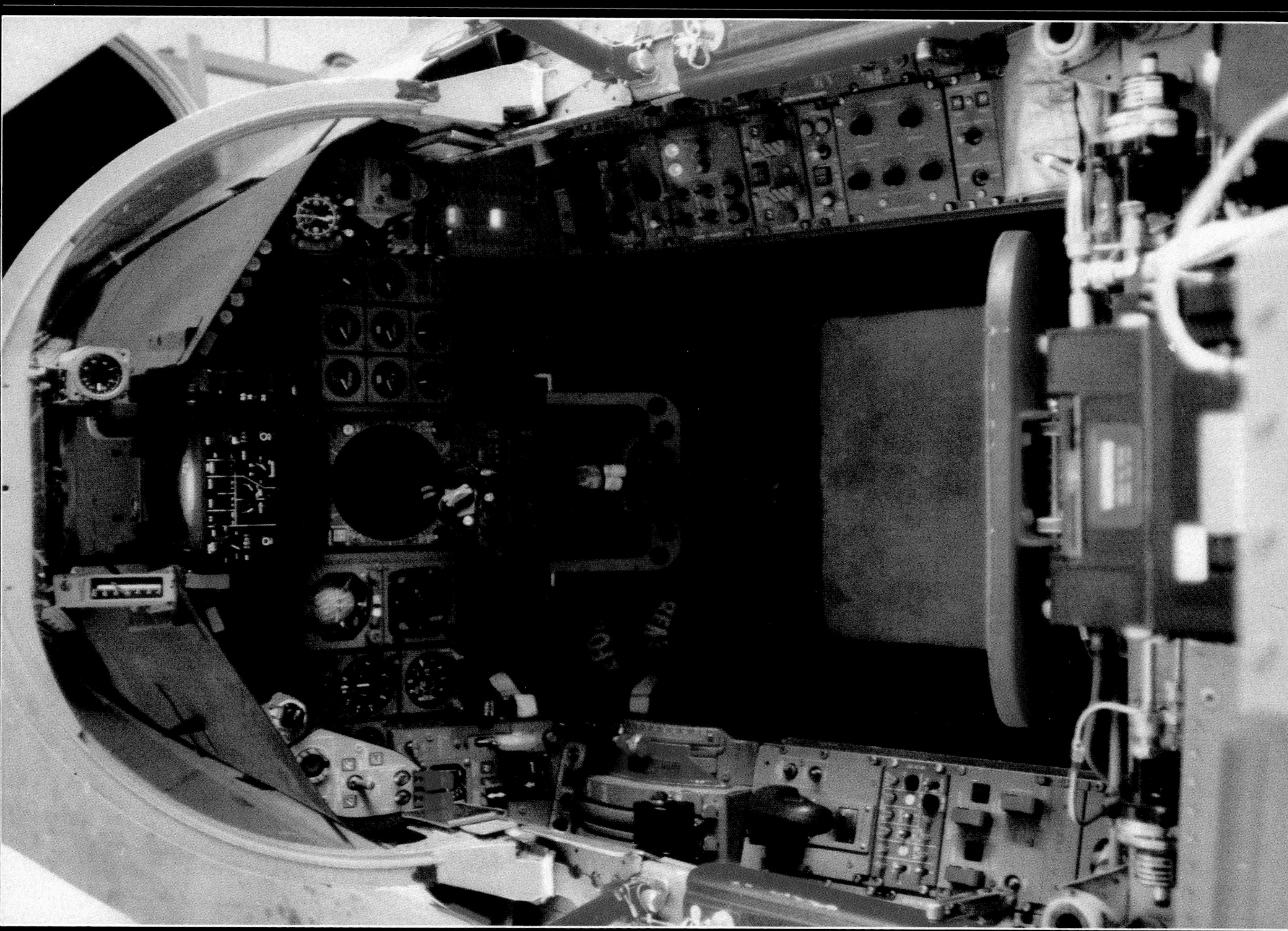

way damaged, and provides an excellent capability against many of the targets invisible to radar or shielded by camouflage.

It is possible that the HUD may be replaced by a system of projecting the data onto the visor mounted on the pilot's helmet, a move which would remove all limitations on field of view. Missiles can already be targeted by helmet-mounted sights; so the helmet HUD is by no means far off. Bio-engineering is also making great strides in parallel with these purely technological developments, and optical control of switches may soon become possible. Voice-activated systems are also being developed, as the cybernetic cockpit moves from the realms of science fiction toward reality.

BELOW Simulated night approach to a carrier.

RIGHT Daylight approach to an airfield in an AV-8B simulator.

BELOW RIGHT Ultimately the military aircraft exists to fight, and the 'Paveway' laser-guided bomb can provide even the most basic type with a pinpoint attack capability.

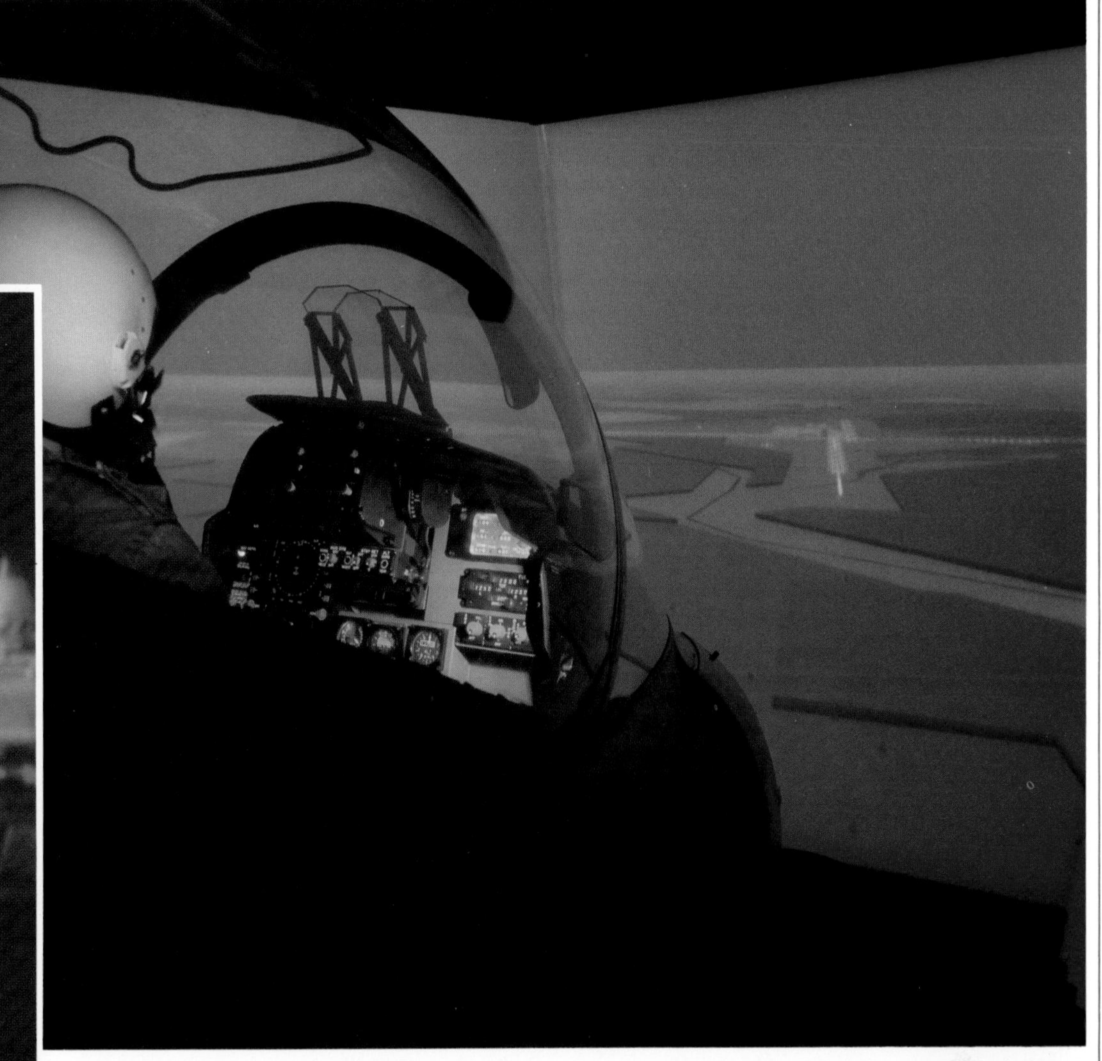

-875 U.S. AIR FORCE

INDEX

H

I

J

K

L

M

N

O

P

Q

R

S

ABOVE LOCKHEED SR-71 'Blackbird'

T

U

V

W

X

Y

PICTURE CREDITS AND ACKNOWLEDGEMENTS

All illustrations for this publication were supplied through Military Archive & Research Services, Lincs. All line artworks from Ravelin Limited. t = top; c = centre; b = bottom; l = left and r = right.

Boeing: p40, 75, 95, 97, 100(b), 101(t), 109(b).
British Aerospace: p38, 39(t), 43(b), 44, 45, 46, 54(t), 63(t), 71(t), 74(t), 80(t), 81(t), 88(c), 89, 100(c), 101(b), 102(b), 105(b), 109(t).
Canadair Ltd: p111(t).
Chemring Ltd: p115(b).
Commonwealth Aircraft Co: p92(t).
Avions M Dassault-Brequet Aviation: p23, 29, 30, 49(b), 50.
Department of Defense: p12, 25, 34(tl), 55(t), 79, 90.
Dornier GmbH: p111(b).
General Dynamics: p18, 20, 21, 22(t), 24(t), 48, 52, 54(b), 83(t), 84(br).
General Electric: p86(t&c).
Grumman Corp: p55(b), 58, 59, 60, 64(b), 65, 88(b), 107, 112, 116(t).
Hughes Aircraft Co: p114(b), 115(t).
Imperial War Museum: p63(b).
Israel Aircraft Industries: p49(t&c), 93(c&b), 94(b).
Lockheed Corp: p11, 26(t), 27, 53, 74(b), 77(b), 113(b).
MARS: p14(b), 96(t), 109(c).
MARS/Niska: p34(b), 51.
Matra: p22(b).
McDonnel Douglas: p8, 9, 17, 19, 41, 42, 43(t), 47, 64(t), 66, 67, 69, 70, 71(b), 78, 84(t&bl), 85, 92(b), 93(t), 100(t), 102(t), 103, 104, 105(t), 106, 117, 118, 119(b), 122, 123(t).
Northrop Corp: p31.
Panavia: p39(b), 120, 121.
Pratt & Whitney: p86(b).
Rockwell International: p16, 28, 32, 33, 56, 57, 96(b).
Rolls-Royce: p80(b), 81(b), 82, 83(b), 87, 88(t), 91.
Royal Netherlands AF: p26(b).
Saab-Scania/Anderson: p36, 37(c&b).
Saab-Scania/Thuresson: p37(t).
SIRPA: p34(tr), 35.
Smiths Industries: p108(t).
Tadiran Ltd: p110.
Texas Instruments: p119(t).
USAF: p10, 15, 24(c&b), 28(c), 76, 77(t), 94(t), 108(b), 113(t), 123(b).
US Navy: p13, 61.
Varo Ins: p14(t).
Vickers Ltd: p62.
Westinghouse Corp: p114(t).